T0213946

Lecture Notes
in Business Information Processing 379

More information about this series at http://www.springer.com/series/7911

Paolo Ceravolo · Maurice van Keulen ·
María Teresa Gómez-López (Eds.)

Data-Driven
Process Discovery
and Analysis

8th IFIP WG 2.6 International Symposium, SIMPDA 2018
Seville, Spain, December 13–14, 2018
and 9th International Symposium, SIMPDA 2019
Bled, Slovenia, September 8, 2019
Revised Selected Papers

 Springer

Editors
Paolo Ceravolo ⓘ
Università degli Studi di Milano
Milan, Italy

Maurice van Keulen ⓘ
University of Twente
Enschede, The Netherlands

María Teresa Gómez-López ⓘ
University of Seville
Seville, Spain

ISSN 1865-1348   ISSN 1865-1356 (electronic)
Lecture Notes in Business Information Processing
ISBN 978-3-030-46632-9   ISBN 978-3-030-46633-6 (eBook)
https://doi.org/10.1007/978-3-030-46633-6

This Springer imprint is published by the registered company Springer Nature Switzerland AG
The registered company address is: Gewerbestrasse 11, 6330 Cham, Switzerland

Preface

The rapid growth at which organizations and businesses process data, managed via information systems, has made available a big variety of information that consequently has created a high demand for making data analytics more effective and valuable. Today, these new data analyzing techniques have to cope with the continuous advancements in the digital transformation course. Blockchain infrastructures bring trusted transactions to interorganizational procedures. The growing maturity level of Artificial Intelligence solutions conveys the integration of various analyzing techniques and cultures. IoT technologies bring traceability to potentially any human-in-the-loop process. The eighth and ninth editions of the International Symposium on Data-driven Process Discovery and Analysis (SIMPDA 2018, 2019) were conceived to offer a forum where researchers from different communities could share their insights in this hot new field. As a symposium, SIMPDA fosters exchanges among academic researchers, industry experts, and a wider audience interested in process discovery and analysis.

Submissions cover theoretical issues related to processing representation, discovery, and analysis, or provide practical and operational examples of their application. To improve the quality of the contributions, the symposium is structured towards fostering discussion and stimulating improvements. Papers are pre-circulated to the authors, who are expected to read them and make ready comments and suggestions. After the event, authors have the opportunity to improve their work by extending the presented results. For this reason, authors of accepted papers were invited to submit extended articles to this post-symposium volume. We received 25 submissions and 6 papers were accepted for publication.

The current selection of papers underlines the most relevant challenges that were identified, and proposes novel solutions for facing these challenges.

In the first paper, "Designing Process-Centric Blockchain-based Architectures: A Case Study in e-voting as a Service," Emanuele Bellini et al. study a solution to put into operation Business Process Management on a Blockchain-based infrastructure to develop a diverse set of execution and monitoring systems on Blockchain, and define appropriate methods for evolution and adaptation.

The second paper, by Berti and Van der Aalst, is titled "Extracting Multiple Viewpoint Models from Relational Databases," and presents an advanced methodology for coping with discovering multiple viewpoints in relational databases collecting event log data.

The third paper by Cancela et al., "Standardizing Process-Data Exploitation by means of a Process-Instance Metamodel," proposes the use of a Business-Process Instance Metamodel as an interface between the applications that produce the data and those which consume the data making their data structures independent and, by consequence, their set up less expensive.

The fourth paper by Hinkka et al., "Exploiting Event Log Event Attributes in RNN Based on Prediction," discusses a method for giving a trade-off between prediction accuracy and training time in RNN predictions over business process cases. This trade-off is achieved by enriching the vectors imputing the RNN using the groups obtained by clustering techniques. This experimental analysis shows that having event attribute clusters encoded into the input vectors outperforms having the actual attribute values in the input vector.

The fifth paper by Martinez-Gil et al., "General Model for Tracking Manufacturing Products Using Graph Databases," presents a case study on product manufacturing where a graph database is exploited to reduce response times in tacking the process execution.

The sixth paper by Rafei et al., "Supporting Confidentiality in Process Mining Using Abstraction and Encryption," presents an approach for supporting data encryption in Process Mining. Using abstraction, the authors hide confidential information in a controlled manner while ensuring that the desired Process Mining results can still be obtained.

We gratefully acknowledge the research community that gathered around the problems related to process data analysis. We would also like to express our deep appreciation for the referees' hard work and dedication. Above all, thanks are due to the authors for submitting the best results of their work to SIMPDA.

We are very grateful to the Università degli Studi di Milano and IFIP for their organizational support.

March 2020

Paolo Ceravolo
Maurice van Keulen
María Teresa Gómez-López

Organization

Chairs

Paolo Ceravolo	Università degli Studi di Milano, Italy
Maurice Van Keulen	University of Twente, The Netherlands
Maria Teresa Gomez Lopez	University of Seville, Spain

Advisory Board

Ernesto Damiani	Università degli Studi di Milano, Italy
Erich Neuhold	University of Vienna, Austria
Philippe Cudré-Mauroux	University of Fribourg, Switzerland
Robert Meersman	Graz University of Technology, Austria
Wilfried Grossmann	University of Vienna, Austria

SIMPDA Award Committee

Maria Teresa Gomez Lopez	University of Seville, Spain
Paolo Ceravolo	Università degli Studi di Milano, Italy

Web and Publicity Chair

Fulvio Frati	Università degli Studi Di Milano, Italy

Program Committee

Alexandra Mazak	University of Vienna, Austria
Robert Singer	FH Joanneum, Austria
Manfred Reichert	University of Ulm, Germany
Schahram Dustdar	Vienna University of Technology, Austria
Helen Balinsky	Hewlett-Packard Laboratories, UK
Valentina Emilia Balas	University of Arad, Romania
Antonio Mana Gomez	Universidad de Málaga, Spain
Karima Boudaoud	École polytechnique de l'université Nice-Sophia Antipolis, France
Jan Mendling	Vienna University of Economics and Business, Austria
Peter Spyns	Flemish Government, Belgium
Mohamed Mosbah	University of Bordeaux, France
Chintan Mrit	University of Twente, The Netherlands
Fabrizio Maria Maggi	University of Tartu, Estonia
Pnina Soffer	University of Haifa, Israel
Matthias Weidlich	Imperial College, UK

Roland Rieke	Fraunhofer Sit, Germany
Edgar Weippl	Vienna University of Technology, Austria
Benoit Depaire	University of Hasselt, Belgium
Angel Jesus	Varela University of Seville, Spain
Luisa Parody	University Loyola Andalucia, Spain
Antonia Azzini	Consorzio C2t, Italy
Jorge Cardoso	University of Coimbra, Spain
Carlos Fernandez-Llatas	Universitat Politècnica de València, Spain
Chiara di Francescomarino	Fondazione Bruno Kessler, Italy
Faiza Bukhsh	University of Twente, The Netherlands
Mirjana Pejifá Bach	University of Zagreb, Croatia
Tamara Quaranta	40labs, Italy
Anna Wilbik	Eindhoven University of Technology, The Netherlands
Yingqian Zhang	Eindhoven University of Technology, The Netherlands
Richard Chbeir	Université de Pau et des Pays de l'Adour, France
Renata Medeiros	Eindhoven University of Technology, The Netherlands
Rabia Maqsood	Università degli Studi di Milano, Italy

Contents

Designing Process-Centric Blockchain-Based Architectures: A Case Study in e-voting as a Service

Emanuele Bellini[1,2(⊠)], Paolo Ceravolo[3], Alessandro Bellini[1], and Ernesto Damiani[2,3]

[1] Mathema s.r.l., 50142 Florence, Italy
{Emanuele.bellini, abel}@mathema.com
[2] Center of Cyber Physical Systems, Khalifa University of Science, Technology and Research, Abu Dhabi 127788, UAE
ernesto.damiani@ku.ac.ae
[3] SesarLAB, University of Milan, Milan, Italy
paolo.ceravolo@unimi.it

Abstract. This article aims at introducing a new process-centric, trusted, configurable and multipurpose electronic voting service based on the blockchain infrastructure. The objective is to design an e-voting service using blockchain able to automatically translate service configuration defined by the end-user into a cloud-based deployable bundle, automating business logic definition, blockchain configuration, and cloud service provider selection. The architecture includes process mining by design in order to optimize process performance and configuration. The article depicts all the components of the architecture and discusses the impact of the proposed solution.

Keywords: Blockchain · Trust e-Vote · Configurability · Automation · Mirror model

1 Introduction

Processes span organizational boundaries, linking together heterogenous information systems. Business Process Management (BPM) has been largely earmarked to address the integration of inter-organizational processes. Coordination and interface of independent systems have been one of the key issues for years [59, 60] but it has been argued that technical integration is simply a precondition of organizational integration that requires trust and comprehension among parties [64]. Recently, the emerging blockchain technology has been acknowledged as the right solution for complementing BPM with support for trustworthy execution of business processes [61, 62]. Putting into operation BPM and Blockchain requires to engage with specific challenges. Following [62] we underline the need for (i) *developing a diverse set of execution and monitoring systems on blockchain*; (ii) *devising new methods for analysis and engineering business processes based on blockchain technology*; (iii) *defining appropriate methods for evolution and adaptation*. All these challenges call for an evolution of the

P. Ceravolo et al. (Eds.): SIMPDA 2018/2019, LNBIP 379, pp. 1–23, 2020.
https://doi.org/10.1007/978-3-030-46633-6_1

technological framework where Blockchain is specified and deployed. As part of this comprehensive view, this paper proposes a vertical study focused on an e-voting as-a-service application.

Different organizations, at different levels, have to organize elections in compliance with specific legal frameworks or standards. Even though we live today in a digital world, voting, in many and diverse contexts, continues to be implemented using the traditional process based on ballot boxes and manual verification. The reason behind the longevity of the traditional process is not without good reasons. It relates to the low transparency and the high vulnerability of distributed computer networks, as testified by the failure of e-voting in Estonia and Australia [7]. Even if its applicability for political pools remains an open challenge, there are a number of cases in daily life in which electronic vote may be successfully applied. Examples include the company's executive elections, where each shareholder is entitled to one vote per share multiplied by the number of executives to be elected. The rule for such voting system may be so rigid ending up to reduce their inclusiveness in some cases (the voting task is performed with a reduced number of voters available) or to affect the operativity of the organization in case of prolonged delay waiting that all the conditions for participation are satisfied. In this respect, electronic voting systems start to be attractive because they can significantly reduce the costs of implementation and verification as well as ease and incentivize the participation of voters thanks to its ubiquity. In recent years, the process of transformation has accelerated, and businesses have entered into the cloud computing era, where virtually turning anything into an online service (XaaS) has become a distinct possibility [25]. In this paper, we introduce HyperVote, a blockchain-based e-voting as a service in the framework of a scalable business model grounding the solution on three pillars: **configurability, automation, transiency.**

HyperVote is built on a Process Centric Architecture (PCA) "*where the entire IT system is conceptualized and centrally organized by the concept of the business process that is supported by the system and the business process component is the central component in the system*" [56]. It is able to support dynamic selection, deployment, and execution of an e-voting process whose requirements are tailored to the user needs. Thus, organizations of any dimension will be able to set-up a secure and transparent voting service using a pay-per-use configurable approach bringing on costs during the limited timeframe of an election. As regards of the service provider, this approach allows optimizing the allocation of resources as the same infrastructure can serve different customers in different time frames.

Electronic voting systems have been fraught with security concerns and controversy. In fact, a voting system has to guarantee three fundamentals requirements: **robustness, uniqueness** of the vote, and **transparency**. In this context, the anonymity that is commonly considered fundamental is here treated as an option in order to include all those cased in which the vote should be evident. Despite several technology and protocols was explicitly designed to meet the three of them simultaneously, this goal is difficult to be achieved in current e-voting systems. In fact, electronic vote is exposed to attacks such as double voting, Sybil attacks, and similar that affect its trustworthiness slowing down its adoption.

The three fundamental requirements of voting are however supported by an emerging technology that more and more observers expect will become a standard in the next future: the Blockchain. In general terms, a Blockchain is an immutable transaction ledger, maintained within a distributed network of peer nodes. Each node is requested to maintain a copy of the ledger by applying transactions that have been validated by a consensus protocol. This model offers great advantages in terms of transparency authenticity, nonrepudiation but imposes severe constraints in terms of performances especially when the identity of a participant must be assessed in conformance to regulations. Thanks to these features, Blockchain technologies are successfully adopted in very diverse domains [57–60]. In this respect, it represents a suitable tool for supporting e-voting, as evidenced by the multiplication of projects focusing on this integration. None of these projects has, however, demonstrated to scale-up, receiving the appropriate acceptance in the market. In fact, there is a large number of examples in which Blockchain is implemented as a static and monolithic technological solution. Startups and ICOs tend to shape their business around one specific solution considered suitable at that time.

This strong dependency between the business case addressed and the type of implementation of the Blockchain exposes those initiatives to a high risk of being run out from business very quickly because of the rapid technology obsolescence that usually characterizes emerging sector as the Distributed Ledger Technology domain. This is also particularly evident when you design and dimension your private Blockchain infrastructure. Parameters as numbers of peers, HW/SW requirements and maintenance, computational performance, and so forth, that have an impact on the sustainability a project in a long run, tend to be neglected in favor of quick delivery of the solution to market. The result is that 92% of all projects fail. Since inception roughly ten years ago, the Blockchain industry has witnessed the launch of 80,000 projects, according to the China Academy of Information and Communications Technology (CAICT). Of them, only 8% are survived while the remaining 92% failed with an average lifespan of 1.22 years [53].

This risk is here mitigated by the adoption of the *Process Thinking* perspective [56], which put the process at the core of the enterprise business process. This is a shift from the convectional *Functions Thinking* to the new process thinking, where functions are enablers to process and process performance and its optimization is more important than optimizing individual functions [56]. Thus in HyperVote, we do not consider Blockchain as a static installation but a function within the Hyper Vote Enterprise Business process that can be dynamically deployed and instantiated according to the service level requested by the end-users for a voting stage. Then, the Blockchain infrastructure is used to execute the HyperVote process in a trustworthy condition while parameters such as number of nodes, computational capacity and service availability offered by the cloud where the node runs, data preservation and so forth, become a matter of service (*process configurability*). This process needs to be deployed in an automatic way, negotiating, each time, the best price offered by the cloud service where the blockchain and the related smart contracts will be deployed (*automation*). Moreover, there is no reason to maintain historical data over its natural lifecycle that is usually defined by the user or by the law. In HyperVote, even if the service is provided using blockchain, the data managed with the blockchain should be decommissioned at some point. In fact, the costs of data preservation should not be sustained by the service provider in a long run, thus the duration of data retention will be defined by the user and its costs computed in the final bill.

The user can decide to keep the blockchain up and running for further analysis and verification after the end of the voting process. This period named *data retention period* (DRP) is part of the service configuration capability of HyperVote whose costs are opportunely computed in the final bill.

In order to manage such a case, HyperVote adopted the strategy to set up a different Hyperledger instantiation for every voting service provided instead of having a single Hyperledger network where multiple services are executed. In this way, as soon as the data retention period is expired, it is possible to automatically turn-off the entire services and concluding the contract with the cloud provider without affecting the other running services. The main aim is to keep under control all the operational costs of the service.

The article is organized as follow: in Sect. 2, a review of the current initiatives on e-voting based on blockchain is presented, in Sect. 3 the end to end verifiability and trust of the e-voting service has been introduced according to the mirror model [41]; Sect. 4 is devoted to defining the service requirements while Sect. 4 provides the reference architecture of HyperVote, Sect. 5 is dedicated to the evaluation and conclusions are provided in Sect. 6.

2 Related Work

There is a vast literature focused on properties of electronic vote protocols [34] and there are a number of functional and security requirements for a robust e-voting service that include transparency, accuracy, auditability, system and data integrity, secrecy/privacy, availability, and distribution of authority presented in [4–6, 13]. These properties are often expressed in terms of formal languages that could be mapped on a technological solution, but this mapping is seldom or only partially provided.

Another mapping that is often neglected is with BPs, meaning real-world operational or normed activities and procedures. The literature focusing on exploiting Blockchain for the electronic vote is a typical example of that. Relevant properties of voting protocols are matched but the integration of these properties with external services, that may be implied by specific organizational or legal constraints, typically is not addressed. We claim that the realization of a fully configurable electronic vote implies the definition of a platform-as-a-service environment addressing the properties of a protocol a three abstraction levels (see Fig. 1):

- Business Process, defining the orchestration of autonomous services;
- Business Logic, defining the behavior of each single service;
- Transaction data, defining the properties transactions have to guarantee.

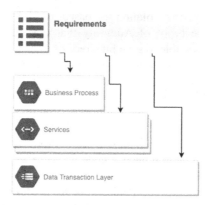

Fig. 1. Logic dependability of the architecture

Several initiatives at commercial level exist as Bitcongress [2], Followmyvote [3], Agora-Voting [43], VoteWarcher, E-VOX [42], and TIVI [4].

Agorà [43] is an end-to-end blockchain-based solution designed for public organizations according to a B2G view. The system is token-based, so that organizations should purchase these tokens for each individual eligible voter. The community of node operators is organized following a hybrid permission/permission-less model and are incentivized to verify election results by the VOTE token. Here the unique factor of scale for the user is the number of the voters and it is not possible to set other parameters such as the level of service required that is assumed standard and independent to any factors. Similarly, BitCongress [2] is a platform that binds together Bitcoin, Counterparty (to create tokens from Bitcoin), Blockchain with smart contracts. For every BitCongress participant, a token is released (called VOTE) which can be sent to a single address, thus not allowing the double vote. Once the elections are over, the tokens are returned to owners. In [27] a new protocol to support that utilizes the Blockchain as a transparent ballot box is provided. In addition to the fundamental properties of a voting system, the protocol allows for a voter to change or cancel one vote, replacing it with another. In [30] the Crypto-voting service is proposed. Crypto voting is built upon SideChain and consists of using two linked blockchains, one-way pegged sidechain, where the main blockchain is used to record eligible voters and record their vote operations, while the pegged one count the votes.

In [32] is designed a voting system based on Multichain. In order to perform a transaction in Multichain, it is needed to identify the address of the node and the related balance from where the asset (vote) will be sent. While sending the asset to the address, the transaction hash was generated carrying the transfer of vote. The balance of the receiving node was incremented by one vote (asset). The transaction becomes a part of the public ledger which shows that it has been mined.

Recently the "as a service" paradigm is emerging also in the blockchain domain in general and in e-voting in particular. For instance, in [31] and [44] blockchain is proposed as a service while in [33] the entire voting system based on blockchain is offered "as a service".

In the first two cases, the solution is based on Ethereum Private Network (EPN) version. In [31] the Proof of Authority that uses the Identity as a stake is introduced. Using EPN is possible to send hundreds of transactions per second onto the blockchain for free, exploiting every aspect of the smart contract to ease the load on the blockchain while keeping the access permissions more tightly controlled, with rights to modify or even read the blockchain state restricted to a few users.

However, the trustworthiness of the solution is strongly dependent on the number of nodes actually deployed. In fact, because of the militated number of nodes in the EPN, it is necessary to reduce the complexity of the Proof of Work (POW) consensus significantly to allow the miner to mine the block in due time. It means that the security and integrity of the private blockchain will not rely anymore on the blockchain itself but on external protections to prevent fraudulent nodes to join your network. Moreover, in POW based private blockchain network it is theoretically possible for a group of nodes to come together to collude and make unauthorized changes to past transactions on the Blockchain. A possible method to prevent this risk is to save a snapshot of the private Blockchain on the public Ethereum network so that a block hash on the private network can be counter-checked against what's written to the public Blockchain.

The second one is based on Hyperledge that is natively designed to support data privacy in a consortium-based network. The trust is granted by design and there is not any PoW or cryptocurrency embedded in the architecture. This means that with a limited number of nodes it is possible to achieve the appropriate level of reliability and security. The voting service including the underling blockchain infrastructure can be tailored by the users that can select a number of parameters as (a) the number of eligible voters, (b) the data retention period, (c) the level of performance (24/7), (d) the tallying method, (e) anonymity, etc. All these parameters concur with the final billing.

The solution proposed in [28] is also based on Hyperledge and stresses the aspects of portability of on different platform of the blockchain-based e-voting service, confirming the capability of Hyperledge project to be used in an on-demand business scenario.

According to this analysis, the paradigm "as-a-service" needs to be exploited in-depth automating most of the steps in setting up an e-voting service such as the blockchain architecture and the development of the related smart contracts in order to implement a fully automated Decentralized Autonomous Organization (DAO) capable to be set up a voting service on demand.

3   End to End Verifiability and Trust in HyperVote

In our system, a voting event is composed by four pillars: a Ballot format, the set of Eligible voters, a Ballot box and the Tallying. In particular, a Ballot can be defined as:

$$B = \langle b_1, b_2, \ldots b_n \rangle$$

Where $B \neq \{\varnothing\}$ b_i is the i-th option (preference) of B and $n = |B|$ represents the cardinality of all the options by which a preference can be expressed. The element b is a boolean variable and represents the preference expressed (e.g. 1 to mark the preference, 0 otherwise). The condition to express the preference maybe several.

The second pillar of a voting event is the vector of eligible voters EV defined as follow:

$$EV = \langle ev_1, ev_2, \ldots ev_m \rangle$$

Where $EV \neq \{\varnothing\}$, ev_i is the i-th eligible voter in the vector EV and m is the cardinality of $|EV|$.

The third pillar is the Ballot Box BB, where all the votes are collected and can be formalized as follow:

$$BB = \langle B_1, B_2, \ldots, B_h \rangle$$

Where h is the cardinality of BB and $h \leq m$.

The last pillar is the Tally process T that starts after the formal conclusion of the ballot and can be defined as follow:

$$T = f(BB)$$

where $f(BB)$ is a generic function in BB (e.g. one of the simplest is the cumulative frequency of $b \in B$) dedicated to counting the votes according to the algorithm defined and decrees the result (R).

A voting event should satisfy some very basic constraints to be considered reliable, such as:

(a) **Eligibility:** it is necessary to verify the eligible voter ev_i before allowing her to express the preference. Typically, this can be obtained by implementing an authentication service.

(b) **1:1 correspondence between B and ev.** An ev_i can express one and only one vote B_i even if multiple preferences are included. Thus, it is necessary to manage the process in such a way to avoid double-voting.

(c) **Privacy (where necessary),** in fact, not only the vote needs to be expressed in a safe and privacy-preserving condition, but it is necessary to pass through a mixing procedure shuffling the BB in order to avoid the possibility to guess the voter on the base of the arrival order of the vote collected. In fact, the temporal dependency between the event of voting by the voter and the event of collecting the vote is a well-known vulnerability that an e-voting system needs to address. In an offline system, the procedures foresee a physical shaking of the BB after the conclusion of the voting process and just before the counting phase.

(d) **Integrity:** Ensure that each vote is recorded as intended and cannot be tampered with in any manner, once recorded (i.e., votes should not be modified, forged or deleted without detection).

(e) **Transparency:** the tallying phase T needs to be managed in such a way the counting is performed with the guarantee that the calculation method is not affecting the final result.

(f) **Verifiable:** Allowing voters and third-party organizations to verify election results.

(g) **Auditable:** Supporting risk-limiting audits that help to assure the accuracy of an election.

All these constraints are a matter of trust and require a reliable solution able to guarantee it. In the present paper, we introduce the concept of End to End (E2E) trust in the e-voting system, i.e. a set features able to guarantee the appropriate level of protection of the votes collected and of the calculation algorithms. Here is where the Blockchain technology came to play.

The introduction of blockchain into a real-life scenario requires an in-depth analysis of the cases to verify (i) if the use of blockchain makes sense and (ii) how to introduce the technology in an appropriate way. In this respect, in [41] the Mirror Model (MM) has been introduced. In MM, three domains are identified: Reality, Representation, and Trust. In particular, *"Reality is where things objectively exist, independently of any efforts made by humans to design or control the Reality, Representation is a model conceived by architects or engineers or scientists to depict the aspects of interest or concern in the Reality and, finally, Trust is a structure of validation of the Representation of Reality whose aim is to make the Representation and its data dependable."* [41]. Through the Trust and Representation domains, we can act indirectly on the Reality to make it more reliable and secure for the benefit of humans and the environment in which they live.

Thus, according to the MM for blockchain adoption on real-life scenarios [41], the e-voting system can be entrusted as depicted in Fig. 2.

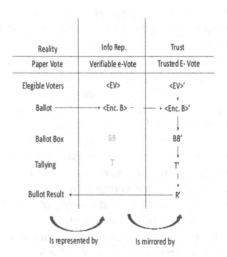

Fig. 2. Mirror model of e-voting scenario

In particular, the *Reality* column includes the main elements composing a paper-based voting event. In the *Information Representation* column, such elements are modeled to implement an e-vote application. Despite it is quite common that in a blockchain-based project only a few parts of the information system created to manage an aspect of the reality are mirrored on the blockchain, in the case of a voting system, the use of the blockchain technology allows different considerations. In fact, because of the risk of introducing weaknesses at the information representation level (e.g. temporal dependencies of the database records link), several elements of the e-vote can be directly implemented within the blockchain infrastructure. However, some precautions must be taken as the generation of blocks has a direct temporal dependency with the vote expression that might be exploited to break the privacy constraint.

To address such issues, e-voting services are adopting cryptographic techniques to secure and end-to-end verifiability while preserving the privacy of the voters and avoiding the discovery of any possible link between a voter and her expressed preferences. One of the most used is *homomorphic encryption* [35, 45] that allows one to operate on ciphertexts without decrypting them. For a voting system, this property allows the encrypted ballots to be counted without leaking any information in the ballot [46, 47]. In fact, individual votes are encrypted votes and then combined to form an encrypted tabulation of all votes which can then be decrypted to produce an election tally that protects voter privacy. By running an open election verifier, anyone can securely confirm that the encrypted votes have been correctly aggregated and that this encrypted tabulation has been correctly decrypted to produce the final tally. This process allows anyone to verify the correct counting of votes by inspecting the public election record while keeping voting records secure. Other approaches include the *zero-knowledge proof* [49, 50] that requires that the voter should convince the authority that his vote is valid by proving that the ballot includes only one legitimate candidate without revealing the candidate information, or the *mix-net* approach [36, 37, 48] that aims to perform a re-encryption over a set of ciphertexts and shuffle the order of those ciphertexts. A mix node only knows the node that it immediately received the message from and the immediate destination to send the shuffled messages to.

However, the simple use of encryption technique, it is not a guarantee of reliability of the entire voting process. For instance, a security vulnerability has compromised 66,000 electronic votes in the New South Wales state in Australia in 2015 [51] even if encryption has been used. To this end, it is necessary to add a mechanism of securing an E2E trustworthy. In this respect, the *BB* implementation at Information Representation level (e-vote) that is usually managed with a database, needs to be avoided in favor of its implementation at the trust level (blockchain) so that the encrypted votes are collected with an untamperable tool. In this sense the mirroring is applied directly on the reality element. Moreover, to avoid attacks on the calculation method, the tallying function needs to be encoded at trust level T' (with a Smart Contract/Chaincode) as well. It will work on the mirrored Ballot Box (*BB'*), and thanks to the Homomorphic encryption the votes will be possible to obtain the results while preserving the privacy. The results of the counting are also managed in the trust level (*R'*) as a transaction in the blockchain that concludes the voting process. This way the entire process is communicated back to reality.

According to this process, the blockchain acts as an E2E trust layer where voters, vote, tallying and the result are stored in an untampered manner allowing further verification and audits.

In this way, it is possible to avoid any cyber-attack focused on votes removing or altering that could occur at the representation level as described in Sect. 2.

HyperVote Enterprise Business Process

In order to define the general business process of HyperVote it is necessary to decompose the entire voting process into self-consistent atomic steps that compose the Business Process. Then, according to MM, it is necessary to identify which of the steps can be implemented in a trustworthy environment using the blockchain. The result of such a decomposition and mapping is provided in Fig. 3. The picture shows that most of the tasks can be implemented in the Trust layer (dotted area), thus using blockchain. In this sense, business logic is defined by the smart contracts selected to implement the Business Process. This smart contract/chaincode can be seen also as a microservices architecture whose atomic services need to be composed to execute the final process.

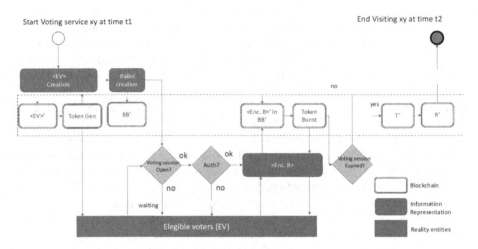

Fig. 3. e-voting business process

Business Process to SOA Translations

The second step includes the translation of the BP into a Service Oriented Architecture model (SOA). SOA [21] is an approach for defining and implementing services over a network linking business and computational resources (principally organizations, applications, and data) to achieve the results requested by service consumers (which can be end-users or other services). The definition, given in [21], now common for Web services and the related architectures, can be also exploited in the blockchain framework for chaincode design. There are different approaches for service modeling, as surveyed in [22] e.g. SOA-RM, SOA-RFA, SOMF, PIM4SOA, SoaML, SOA ontology, and SOMA. In the present work, we adopted SOMF since it provides a formal method of service identification at different levels of abstraction, from holistic to detailed, including meta-model concept and specific notation [22–24]. More specifically, we extended the SOMF notation introducing the Trusted Atomic Service (TAS) (see Fig. 4). A TAS is an irreducible service resulted from the decomposition of the tasks executed on the blockchain. It is implemented by a chaincode and, according to the MM, may represent a mirrored Atomic Service. According to the extended SOMF, the BP is translated into the service architecture depicted in Fig. 5.

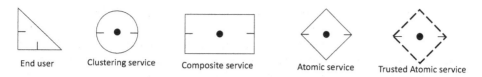

End user   Clustering service   Composite service   Atomic service   Trusted Atomic service

Fig. 4. SOMF notation plus trusted atomic service

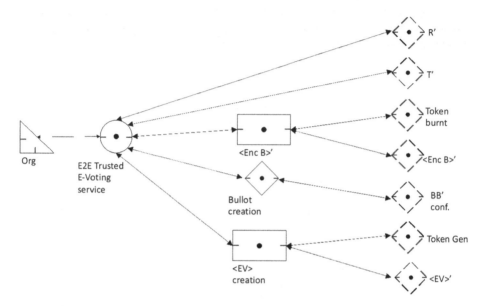

Fig. 5. SOMF-based HyperVote business process translation

Process Analytics Flow

The subject of measurement in process analytics are *events* that occur during the process execution. An event represents the change of state of objects such as data, activities, parameters, users, and so forth, within a given context [55]. For instance, log-in/log-out, process start/end, CRUD operations are *events* that need to be detected and then analysed to monitor system performance and compliance against business KPIs. Because of the complexity of the e-voting, it is useful to define a model organizing the steps in the knowledge flow. In HyperVote there are three classes of *events* to be detected: *process configuration events*, *process execution events*, and *compliance check events*. Process configurations events are recorded each time one of the parameters that the user can manipulate and customise is altered. Tracking these changes generates data that can be used to understand user intention and preferences. In addition, Cloud and Hyperledger Fabric infrastructures should be monitored and assessed against their performances. For instance, the cloud service on which the Fabric is deployed can be evaluated against parameters such as *availability, reliability, response time, security, throughout, capacity, scalability,* and *latency*. Similarly, the chaincodes executed in the blockchain can be analysed against their memory and CPU consumption. However, such data are interrelated with complex and sometimes not linear dependencies. In fact, the performance of the blockchain is only partially affected by the computational infrastructure on which it is deployed, the selected configuration (e.g. number of nodes) and the chaincode implementation style/logic also play a crucial role. In other words, to optimize the HyperVote service for both user and service provider (e.g. MAX(*Profit*), MIN (*SLA deviation*)), it is necessary to continuously collect through an ETL mechanism. The absence of logging libraries in blockchain frameworks makes event logs generation a project critical task requesting dedicated development effort. The technological scenario is however in evolution with initial logging solutions devoted to blockchain technology [63] (Fig. 6).

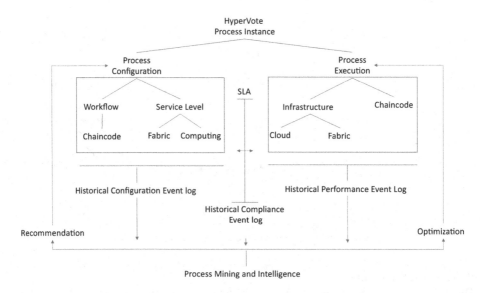

Fig. 6. HyperVote process analysis flow

4   HyperVote Process Centric Architecture

The concept of Process Centric Architecture (PCA) to create software systems that enable business process has been introduced in [56]. The objective of PCA is to support the creation of architectures able to execute business processes of an enterprise. Figure 7 depicts the architecture with the main components and their relationship with the Hypervote Cloud Bundle (CB), that will be used in the Cloud Deployer Orchestrator for the process execution. The HyperVote CB is an XML based data structure containing all the information required by the execution environment to deploy, execute, account, and monitor the service. The HyperVote CB binds a Workflow Model with the concrete Atomic Services (chaincode) and Hyperledger Network structure and Cloud Infrastructures that will be invoked and selected by the workflow, respectively, when executed. The HyperVote CB is structured as follow:

- Section 1 - Voting Process parameters. This section includes all the parameters defining the Voting Event such as Date time for start/end, and the parameters defined in Table 1.
- Section 2 - Voting Process model. Such a model is an executable BPMN 2.0 file already configured with all the technical details required to be executed by a BPM Engine. It might contain invocation to atomic service selected from the Smart Contracts included in the process.
- Section 3 – Trusted Business Logic. This section includes the reference to the selected chaincodes that implement each step in the BPMN model.
- Section 4 – Fabric Architecture parameters. This section includes the definition of the number of peers to be deployed, which will be configured as endorser, i.e. the Certification Authority and the Order. It is important to remark that a core architecture is defined and includes three peer nodes, one Order and one CA that are executed within the service provider.
- Section 5 – Cloud Services. This section includes information about the cloud provider, the services to be used, and the negotiated best SLA on the Cloud Providers Catalog. The parameters can conform to a combination of WS-Agreement (WS-Agreement 2015) and OWL-Q [51]. The allocation-related information is represented according to CAMEL [29], a multipurpose DSL that comprises existing DSLs such as CloudML.

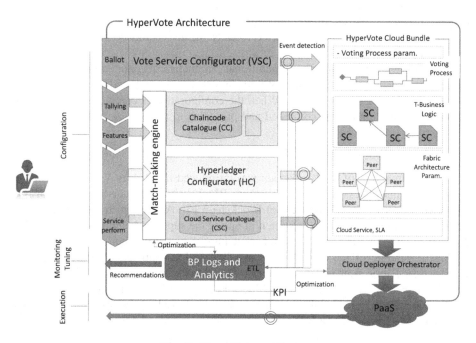

Fig. 7. HyperVote architecture

As defined in PCA, the Business Process components are here represented by the Vote Service Configurator (VSC) and the related Voting Process. The Business Functions components are represented by the Chaincode Catalog and the related Trusted Business Logic included in the CB as well as the Match-making engine. The User Interface is represented by VSC frontend and the HyperVote end-user application for voting.

4.1 Vote Service Configurator (VSC)

The Vote Service Configurator (VSC) is the interface a customer can use to specify the properties required by the system. Each selection is reflected in a change in the Service Level Agreement and the Pricing Scheme to be signed by the parties. In particular, the configuration is supported by a wizard organized in five steps reflecting the HyperVote Cloud Bundle organization. Thus, each step is devoted to trigger a platform component that manages the related section in the bundle.

Through the VSC it is possible to select the appropriate tallying mechanism searching into the Chaincode Catalogue. In fact, there are a relevant number of different tallying process. For instance, in a Ranked Choice Voting (RCV) mechanism, voters can rank as many candidates as they want in order of choice. Candidates do best when they attract a strong core of first-choice support while also reaching out for second and even third choices. When used as an "instant runoff" to elect a single candidate like a mayor or a governor, RCV helps to elect a candidate that better reflects the support of a majority of voters. Other mechanisms are related to exclusive vote so that the voter can express a mutually exclusive preference for only one candidate. In this regard, during the configuration phase, the user will be looking for the chaincode that better reflects the intent of the user.

VSC allows the user to specify additional parameters, as defined in Table 1, and supports the selection of the Quality of Service that will be translated into an SLA once the negotiation phase is concluded. When the configuration is completed, the prize of the e-voting solution can be computed and presented to the user.

4.2 Chaincode Catalogue (CC)

The programmability of the Blockchain was introduced in 2014 by Vitalik Buterin [5] with the permissionless (or open) Ethereum Blockchain, and later by IBM with the Hyperledger permissioned (or private) Blockchain [6]. According to Hyperledge definition, *Chaincode* is a software (written in one of the supported languages such as Go, Node.js or Java) that defines assets and related transactions; in other words, it contains the business logic of the system. It is installed and instantiated onto a network of Hyperledger Fabric peer nodes, enabling transactions with that network's shared ledger. A chaincode typically handles business logic agreed to by members of the network, so it may be considered as a "smart contract". In order to maximize the reuse of the chaincode when similar voting services are required, HyperVote implements a catalog where it is possible to retrieve the chaincode that match the configuration selected by the user. In particular, the chaincodes implementing the tallying process are the most critical. In fact, the users need to select the right tallying mechanism and in case parametrize it.

4.3 Hyperledger Configurator (HC)

The Hyperledger Fabric is a permissioned consortium driven private blockchain network [26], built from the peers owned and contributed by the different organizations that are members of the network. The network exists because organizations contribute their individual resources to the collective network. Peers are a fundamental element of the network because they host ledgers and smart contracts. A peer executes chaincode, accesses ledger data, endorses transactions, and interfaces with applications. Peers have an identity (digital certificate) assigned by a Certificate Authority to authenticate member identity and roles. The validators are known, so any risk of a 50 + 1 attack arising from some miner collusion applies.

Because of that, the Blockchain Configurator is devoted to translating the configuration choices into parameters automating the infrastructure deployment.

Usually, the Hyperledger configuration is a time-consuming task that needs to be tailored on the base of the actual business requirements. For instance, if the number of running chaincodes on the peers exceeds the real need of duplication (e.g. not all the peers needs to host a chaincode in the network), this leads to an increment of the costs of operating the service without, however, obtaining a benefit for either the end-user or the service manager. However, such kind of recommendation is related to consulting activities that could be costly and require effort that might be difficult to estimate. To this end, the blockchain configuration will be a matter of service. Thus, HyperVote may provide some recommendations (or a default configuration), but the final choice remains on the user side. Expert users will be more capable to reduce the costs of the service optimizing the system configuration. Examples of parameters that can be configured by the users are:

(a) *# of running peers*: there is a minimum of three peers that are managed by the service provider.
(b) *chaincode level of distribution*: how many peers will run the chaincode in parallel. This parameter may affect the performance of the service. In fact, having multiple peers executing the chaincode reinforces the robustness of the process but, in turn, the speed to complete the process is reduced.
(c) *# of endorser node*: this parameter defines how many nodes will be set as endorsers. The number of endorsers affects the computational speed but, on the other hand, the process may result in a more reliable data processing.

4.4 Cloud Service Catalogue (CSC)

This component is organised according to the results of the CloudSocket project [54]. The Cloud Service Catalogue is defined by two sub-components: the Cloud Provider

Registry (CPR) and the Atomic Service Registry (ASR). The CPR is responsible to store and describe the different cloud provider configurations (local or remotes) as the login, front-end, the definition of the APIs. This registry is related to the complete management information (instead of being a simple configuration description); only the cloud providers included in the catalog will be used in the HyperVote Cloud Bundle. The ASR describes the cloud services that a Service Provider offers. There are three essential components to the CSC:

- The data model (in a self-describing JSON format) holding the information for the Service Catalogue to describe standard and extended attributes of Service Providers and Cloud Service offerings.
- The standard set of structures to describe the Service Provider and Service Provider's products and services.
- The API to interact with the CSC.

The different workflows use them to be part of the service tasks in the business process definition. Hence, these descriptions have to include both functional descriptions, business details and technical specifications (as the interface description). Besides, this is required certainly in case of adaptation reasons, as we might need to substitute one service with another one. Additionally, it can also be used to draw additional information about a particular service, which could then facilitate its (adaptive) execution [54].

4.5   Cloud Deployer Orchestrator (CDO)

This component is devoted to executing the HyperVote Cloud Bundle. In particular, it interacts with the cloud service where the Hyperledge infrastructure and the chaincodes will be deployed. It will set up all the connections and configuration and will execute a rule-based workflow engine to monitor the steps of the service. Moreover, it will continue to monitor the SLA and provides alerts in case of anomalies reducing the effort to manage running services.

4.6   Match-Making Engine (MME)

The MME component implements the PCA Business Rule layer, where the service provider business rules manage the entire life cycle of the service.

4.7   Business Process Logs and Analytics

Finding a near-optimal configuration for highly configurable systems, such as for instance Software Product Lines, requires dealing with exponential space complexity [8]. This relates to the size of the combinations to be considered (the power set of the assessed features) but also to the fact the interactions between features introduce performance dependencies. Thus, for determining performances, it is not possible to simply aggregate the contributes provided by every single feature. Software systems require to be adapted to a variety of environments, constraints, or objectives. For this reason, modern software is designed to support different configurations and a key goal is reasoning about the performances lead by a specific configuration. A naive approach may consider the system simply as a black box and produce a set of observations for measuring the performances obtained by the system in specific configurations. This approach is however infeasible in any context where the number of configurations is so high that their combinations become intractable.

To make performance prediction practicable two general strategies have been followed, even if, in practice, several works apply a combination of the two.

- **Sampling-centric.** The goal is generating a model from a restrict set of samples keeping, at the same time, the prediction accuracy significantly. Random sampling is regarded as the standard operating procedure [9, 10], however, the true sampling approach should consider only valid configurations. Filtering out invalid configurations is a possible answer [11] but this solution does not avoid generating the entire set of configurations [12]. Encoding configurations in a feature model that using propositional formulas can reveal inconsistent configurations, i.e. conjunctions of features in the model, avoiding exploring the entire configuration space [8]. Statistical procedures have been adopted in order to sampling using the most significant variance [14], if necessary, in conjunction with other variables, such as acquisition cost, for instance [15]. The prediction power of a method can be improved by exploiting the information that is gained from the system itself. Machine Learning has been often used in this respect as it allows an iterative approach with a training set used to create the model and testing set to assess the sample and increase it if the achieve accuracy is low [16].
- **Model-centric.** Measuring the performances of a real system is typically time-consuming, costly and implies risks of damaging the system itself. For this reason, model-driven approaches may be preferred. For example, performance annotated software architecture models (e.g. UML2) can be transformed into stochastic models (e.g. Petri Nets) to then apply analytical or simulation-based optimization methods. A comparative assessment of different approaches with the trade-offs they entail is proposed in [17]. Feature Models can be exploited to represent the interdependencies between features in configurations restricting the sampling space to valid configurations [18] or identifying the minimum set of features required to determine a performance [18]. Clearly, the drawback is that not all inconsistencies are known a priori.

Our approach implements a model-centric view as it guarantees a fast application at the different specification levels handled by our architecture. An example of the parameters that can be defined provided in Table 1.

Table 1. Service configuration parameter

Param	Value	Description
Secrecy/Privacy	Yes/no	This option consists of allowing end-users in disclosing their vote when it is required by the voting system (e.g. qualified vote). In any case, the vote is collected and managed as encrypted
Tallying methods	<list>	This parameter allows the end-user to select the tallying method suitable for the event. A pre-filled list is presented and includes simple and complex calculation methods In case, none of the presented methods works for the case, a wizard for self-defined tallying, method is provided
Under-voting alert	Yes/no	The vote organizer may receive a warning of not voting. However, the system must not prevent under-voting
Receipt	Yes/no	The system may issue a receipt to the voter if and only if it can be ensured that vote-coercion and vote-selling are prevented, so that he may verify his vote at any time and also contend, if necessary
Distribution of Authority	#external running peers	The minimum number of peers to execute Hyperledger is 3. These nodes are executed within the Hyperledger Service; however, a number of external nodes can be instantiated for transparency allowing process inspection to the voter organization(s)
Data retention period	#Days	This feature represents the number of days the instantiated blockchain needs to be kept up and running for any audit and evidence analysis
Reporting	- Simple - Complete - Analytics	This parameter defines the amount of information to be sent back to the user after the end of the tallying process. The simple option reports the winner and ranks the other options; the complete option includes also statistics such as the number of voters, Analytics option add some flow analysis (timeslot based), etc.

To collect events, this component uses an ETL (Extract Transform and Load) process (e.g., Penthao Kettle) invoking a web service via HTTP Post. In most cases, the real-time data are pushed directly into the mapping process to feed a temporary SQL store. They are typically streamed both into a traditional SQL store and then converted into triples in the RDF (using the Karma tool) final store and in a NoSQL database (HBase).

5   HyperVote Evaluation

HyperVote needs to be considered as an agile and cost-effective solution to run a reliable and ubiquitous e-voting service. In order to evaluate the benefit of the proposed approach, we identify the following criteria derived by [38]:

(a) *Non-regression property (NRP)*: the validity of the model developed should not be affected by the addition of new elements (e.g. a new tallying function).

(b) *Rebound effect (RE)* [39]: the increased efficiency in performing the voting service creates an additional demand for new services. It is a proxy indicator of the achieved efficiency.

(c) *Value of the Service* (VoS) [40] defined as $VoS = \mathcal{F}\{t, c\}$ where t is time needed to obtain the result R after the vote collection; c is the costs to run the service.

(d) *Accuracy (Acc)*: it is related to the capacity of the system to produce a result that is the exact expression with a low number of votes considered not valid or lost.

(e) *Effort*: this criterion refers to the total effort in days needed to run an election service.

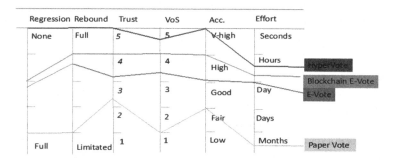

Fig. 8. HyperVote evaluation

According to the evaluation framework, HyperVote is absolutely not regressive, since changes in the Service configuration (e.g. a different tallying algorithm), does not invalidate the rest of the Service process. In fact, configurability is one of the main features of the system.

On the contrary, for paper-based/offline voting system, every change (e.g. preference expression mechanism) requires the entire re-execution of the simulation and the results. Blockchain-based voting service, as discussed in Sect. 2, cannot be entirely not regressive because of the use of the Distributed Ledger Technology as a monolith that reduces the possibility of extension and technology adaptation in case of changing conditions (e.g. technology obsolescence).

The possibility to have results in a cost-effective, trusted and verifiable/auditable way triggers the so-called rebound effect [39] in HyperVote that can be expressed in a request of even more configuration levels. Rebound effect is also present in blockchain-based voting system and in simple e-Vote service because of the use of technology. However, the effect is also dampened since the services do not easily scale in presence of a high number of voters (e.g. ELIGO e-voting service requires a project-based approach for more than 2000 voters involving a consulting phase that increases the costs [19]). Because of automation, the VoS is maximum in HyperVote. It is also significant for e-voting services. The combination of E2E verifiability with the E2E trust assigns the maximum value for the accuracy to Hypervote. The simple use of blockchain increases the accuracy in e-voting services, while for paper vote, the voting process mechanisms are able to guarantee just a fair level. Finally, the effort required to configure lunch and obtain the results of the consultation in HyperVote is quantifiable in terms of hours. In the case of a relevant number of voters, in e-voting service, the effort required might be increased. The difference between the paper-based voting process is incomparable. The result of this evaluation is depicted in Fig. 8.

6 Conclusion

Our proposal aims at developing an e-Vote-as-a-Service based on Blockchain overcoming the limitations of the current projects using a cloud-based approach. Even if a number of cloud providers such as IBM and Oracle are offering ready-to-use blockchain installation on the Cloud with a fee based on the number of transactions, the challenge of a dynamic, cross-platform and on-demand system configuration and optimization based on end user's business requirements remains. HyperVote aimed at:

- keeping the operational costs (transaction, maintenance, etc.) is entirely under control of the organizations that provide the service. In particular, since Fabric is based on certified nodes, it is possible to have high performance and scalability respect to the permission less Blockchains,
- combining Privacy by Homomorphic encryption and Trust blockchain allowing tallying process on encrypted votes without the disclosure of the voter identity,
- modeling complex Business Logic with chaincode treated as an atomic service,
- enabling high configurability of the system and cross-platform agile portability.

The proposed architecture aims also to propose a new perspective for the blockchain-based application moving from a static to a dynamic and configurable view to optimize costs and service performance in view of long-term sustainability of the service provided.

References

1. Hardwick, F.S., Gioulis, A., Akram, R.N., Markantonakis, K.: E-voting with blockchain: an e-voting protocol with decentralisation and voter privacy. In: IEEE Conference on Internet of Things, Green Computing and Communications, Cyber, Physical and Social Computing, Smart Data, Blockchain, Computer and Information Technology, Congress on Cybermatics (2018)
2. BitCongress: Control the world from your phone. http://www.bitcongress.org/BitCongress Whitepaper.pdf
3. FollowMyVote.com: Technical report (2017). https://followmyvote.com
4. Tivi - verifiable voting: Accessible, anytime, anywhere, TIVI, Technical report (2017). https://tivi.io
5. Wang, K.-H., Mondal, S.K., Chan, K., Xie, X.: A review of contemporary e-voting: requirements, technology, systems and usability. Data Sci. Pattern Recognit. 1(1), 31–47 (2017)

6. Gritzalis, D.A.: Principles and requirements for a secure e-voting system. Comput. Secur. **21** (6), 539–556 (2002)
7. Halderman, J.A.: Practical attacks on real-world e-voting. In: Real-World Electronic Voting, pp. 159–186. Auerbach Publications (2016)
8. Oh, J., Batory, D., Myers, M., Siegmund, N.: Finding near-optimal configurations in product lines by random sampling. In: Proceedings of the 2017 11th Joint Meeting on Foundations of Software Engineering, pp. 61–71. ACM (2017)
9. Zhang, Y., Guo, J., Blais, E., Czarnecki, K.: Performance prediction of configurable software systems by fourier learning (t). In: 2015 30th IEEE/ACM International Conference on Automated Software Engineering (ASE), pp. 365–373. IEEE (2015)
10. Siegmund, N., Grebhahn, A., Apel, S., Kästner, C.: Performance influence models for highly configurable systems. In: Proceedings of the 2015 10th Joint Meeting on Foundations of Software Engineering, pp. 284–294. ACM (2015)
11. Sayyad, A.S., Menzies, T., Ammar, H.: On the value of user preferences in search-based software engineering: a case study in software product lines. In: Proceedings of the 2013 International Conference on Software Engineering, pp. 492–501. IEEE Press (2013)
12. Henard, C., Papadakis, M., Harman, M., LeTraon, Y.: Combining multiobjective search and constraint solving for configuring large software product lines. In: Proceedings of the 37th International Conference on Software Engineering, vol. 1, pp. 517–528. IEEE Press (2015)
13. Anane, R., Freeland, R., Theodoropoulos, G.: E-voting requirements and implementation. In: The 9th IEEE CEC/EEE 2007, pp. 382–392. IEEE (2007)
14. Guo, J., Czarnecki, K., Apel, S., Siegmund, N., Wasowski, A.: Variability-aware performance prediction: a statistical learning approach. In: 2013 IEEE/ACM 28th International Conference on Automated Software Engineering (ASE), pp. 301–311. IEEE (2013)
15. Sarkar, A., Guo, J., Siegmund, N., Apel, S., Czarnecki, K.: Cost efficient sampling for performance prediction of configurable systems (t). In: 2015 30th IEEE/ACM International Conference on Automated Software Engineering (ASE), pp. 342–352. IEEE (2015)
16. Jamshidi, P., Casale, G.: An uncertainty-aware approach to optimal configuration of stream processing systems. In: 2016 IEEE 24th International Symposium on Modeling, Analysis and Simulation of Computer and Telecommunication Systems (MASCOTS), pp. 39–48. IEEE (2016)
17. Brosig, F., Meier, P., Becker, S., Koziolek, A., Koziolek, H., Kounev, S.: Quantitative evaluation of model-driven performance analysis and simulation of component-based architectures. IEEE Trans. Softw. Eng. **41**(2), 157175 (2015)
18. Schröter, R., Krieter, S., Thüm, T., Benduhn, F., Saake, G.: Feature-model interfaces: the highway to compositional analyses of highly-configurable systems. In: Proceedings of the 38th International Conference on Software Engineering, pp. 667–678. ACM (2016)
19. https://www.eligo.social/
20. Nuseibeh, B., Easterbrook, S.: Requirements engineering: a roadmap ICSE'00. In: Proceedings of the Conference on the Future of Software Engineering, pp. 35–46 (2000)
21. Bell, M.: Introduction to service-oriented modeling. In: Service-Oriented Modeling: Service Analysis, Design, and Architecture. Wiley (2008). ISBN 978-0-470-14111-3
22. Mohammadi, M., Mukhtar, M.: A review of SOA modeling approaches for enterprise information systems. Elsevier Procedia Technol. **11**, 794–800 (2013)
23. Bellini, E., et al.: Interoperability knowledge base for persistent identifiers interoperability framework. In: IEEE Eighth International Conference on Signal Image Technology and Internet Based Systems, SITIS 2012r, vol. 6395182, pp. 868–875 (2012)
24. Truyen, F.: Enacting the Service Oriented Modeling Framework (SOMF) using Enterprise Architect (2011)

25. Duan, Y., Cao, Y., Sun, X.: Various AAS of everything as a service. In: 2015 16th IEEE/ACIS International Conference Software Engineering, Artificial Intelligence, Networking and Parallel/Distributed Computing (SNPD), p. 16, June 2015
26. Buterin, V.: On public and private Blockchains (2015)
27. Hardwick, F.S., Gioulis, A., Akram, R.N., Markantonakis, K.: E-voting with blockchain: an e-voting protocol with decentralisation and voter privacy. In: IEEE Confs on Internet of Things, Green Computing and Communications, Cyber, Physical and Social Computing, Smart Data, Blockchain, Computer and Information Technology, Congress on Cybermatics (2018)
28. Yu, B., Liu, J.K., Sakzad, A., Nepal, S., Steinfeld, R., Rimba, P., Au, M.H.: Platform-independent secure blockchain-based voting system. In: Chen, L., Manulis, M., Schneider, S. (eds.) ISC 2018. LNCS, vol. 11060, pp. 369–386. Springer, Cham (2018). https://doi.org/10.1007/978-3-319-99136-8_20
29. Rossini, A., et al.: D2.1.2– CloudML Implementation Documentation. Passage project deliverable, April 2014
30. Fusco, F., Lunesu, M.I., Pani, F.E., Pinna, A.: Crypto-voting, a blockchain based e-voting system. In: 10th International Conference on Knowledge Management and Information Sharing (2018)
31. Hjlmarsson, F., Hreiarsson, G.K., Hamdaqa, M., Hjlmtysson, G.: Blockchain-based e-voting system. In: IEEE 11th International Conference on Cloud Computing (2018)
32. Khan, K.M., Arshad, J., Khan, M.M.: Secure digital voting system based on blockchain technology. Int. J. Electron. Gov. Res. **14**(1) (2018)
33. Bellini E., Ceravolo, P., Damiani, E.: Blockchain-based e-Vote-as-a-Service. In: IEEE CLOUD Conference (2019)
34. Delaune, S., Kremer, S., Ryan, M.: Verifying privacy-type properties of electronic voting protocols. J. Comput. Secur. **17**(4), 435–487 (2009)
35. Aziz, A., Qunoo, H., Samra, A.A.: Using homomorphic cryptographic solutions on e-voting systems. Int. J. Comput. Netw. Inf. Secur. **10**(1), 44–59 (2015)
36. Sako, K., Kilian, J.: Receipt-free mix-type voting scheme. In: Guillou, L.C., Quisquater, J.-J. (eds.) EUROCRYPT 1995. LNCS, vol. 921, pp. 393–403. Springer, Heidelberg (1995). https://doi.org/10.1007/3-540-49264-X_32
37. Catalano, D., Jakobsson, M., Juels, A.: Coercion - resistant electronic elections. In: Proceedings of the 2005 ACM Workshop on Privacy in the Electronic Society, pp. 61–70. ACM (2005)
38. Bellini, E., Coconea, L., Nesi, P.: A functional resonance analysis method driven resilience quantification for socio-technical systems. IEEE Syst. J. (2019). https://doi.org/10.1109/JSYST.2019.2905713
39. Gossart, C.: Rebound effects and ICT: a review of the literature. In: Hilty, L.M., Aebischer, B. (eds.) ICT Innovations for Sustainability. AISC, vol. 310, pp. 435–448. Springer, Cham (2015). https://doi.org/10.1007/978-3-319-09228-7_26
40. Sajko, M., Rabuzin, K., Bača, M.: How to calculate information value for effective security risk assessment. J. Inf. Organ. Sci. **30**(2)
41. Bellini, A., Bellini, E., Gherardelli, M., Pirri, F.: Enhancing IoT data dependability through a blockchain mirror model. Future Internet **11**, 117 (2019)
42. http://e-vox.org/
43. https://www.agora.vote/
44. Hjálmarsson, F.Þ., Hreiðarsson, G.K., Hamdaqa, M., Hjálmtýsson, G.: Blockchain-based e-voting system. In: 11th IEEE Conference on Cloud Computing (2018). https://doi.org/10.1109/cloud.2018.00151
45. Gentry, C.: A fully homomorphic encryption scheme. Stanford University (2009)

46. Cramer, R., Gennaro, R., Schoenmakers, B.: A secure and optimally efficient multi-authority election scheme. Trans. Emerg. Telecommun. Technol. **8**(5), 481–490 (1997)
47. Katz, J., Myers, S., Ostrovsky, R.: Cryptographic counters and applications to electronic voting. In: Pfitzmann, B. (ed.) EUROCRYPT 2001. LNCS, vol. 2045, pp. 78–92. Springer, Heidelberg (2001). https://doi.org/10.1007/3-540-44987-6_6
48. Chaum, D.L.: Untraceable electronic mail, return addresses, and digital pseudonyms. Commun. ACM **24**(2), 84–90 (1981)
49. Chow, S.S., Liu, J.K., Wong, D.S.: Robust receipt-free election system with ballot secrecy and verifiability. In: NDSS, vol. 8, pp. 81–94 (2008)
50. Neff, C.A.: A verifiable secret shuffle and its application to e-voting. In: Proceedings of the 8th ACM conference on Computer and Communications Security, pp. 116–125. ACM (2001)
51. NSW election result could be challenged over ivote security flaw (2015). https://www.theguardian.com/australia-news/2015/mar/23/nsw-election-result-could-bechallenged-over-ivote-security-flaw
52. Kritikos, K., Plexousakis, D.: Semantic QoS metric matching in ECOWS. IEEE Comput. Soc. **2006**, 265–274 (2006)
53. https://hacked.com/92-of-blockchain-projects-fail-according-to-new-chinese-study/retrieved. Accessed 19 June 2019
54. Woitsch, R.: D4.1 First CloudSocket Architecture (2015). https://site.cloudsocket.eu/documents/251273/350509/CloudSocket_D4.1_First-CloudSocket-Architecture-v1.0_BOC-20150831-FINAL.pdf/0bdd6c7b-a349-47ab-bed9-631e567365d8?download=true. Accessed 19 June 2019
55. zur Muehlen, M., Shapiro, R.: Business Process Analytics. In: Rosemann, M., vom Brocke, J. (eds.) Handbook on Business Process Management, vol. 2, pp. 137–157. Springer, Berlin (2010). https://doi.org/10.1007/978-3-642-01982-1_7
56. Seshan, P.: Process-Centric Architecture for Enterprise Software Systems. Auerbach Publications, Boca Raton (2010)
57. Bellini, E.: A blockchain based trusted persistent identifier system for big data in science. Found. Comput. Decis. Sci. **44**(4), 351–377 (2019)
58. Brotsis, S., et al.: Blockchain solutions for forensic evidence preservation in IoT environments. In: Proceedings of the 2019 IEEE Conference on Network Softwarization: Unleashing the Power of Network Softwarization, NetSoft, June 2019, Article number 8806675, pp. 110–114 (2019)
59. Kettinger, W.J., Grover, V.: Toward a theory of business process change management. J. Manag. Inf. Syst. **12**(1), 9–30 (1995)
60. Vukšić, V.B., Bach, M.P., Popovič, A.: Supporting performance management with business process management and business intelligence: a case analysis of integration and orchestration. Int. J. Inf. Manag. **33**(4), 613–619 (2013)
61. Weber, I., Xu, X., Riveret, R., Governatori, G., Ponomarev, A., Mendling, J.: Untrusted business process monitoring and execution using blockchain. In: La Rosa, M., Loos, P., Pastor, O. (eds.) BPM 2016. LNCS, vol. 9850, pp. 329–347. Springer, Cham (2016). https://doi.org/10.1007/978-3-319-45348-4_19
62. Mendling, J., et al.: Blockchains for business process management-challenges and opportunities. ACM Trans. Manag. Inf. Syst. (TMIS) **9**(1), 4 (2018)
63. Klinkmüller, C., Weber, I., Ponomarev, A., Tran, A.B., van der Aalst, W.: Efficient logging for blockchain applications (2020). https://arxiv.org/abs/2001.10281. Accessed 28 Jan 2020
64. Ghosh, A., Fedorowicz, J.: The role of trust in supply chain governance. Bus. Process Manag. J. **14**, 453–470 (2008)

Extracting Multiple Viewpoint Models from Relational Databases

Alessandro Berti$^{(\boxtimes)}$(iD) and Wil van der Aalst(iD)

Process and Data Science Department, Lehrstuhl fur Informatik 9,
RWTH Aachen University, 52074 Aachen, Germany
a.berti@pads.rwth-aachen.de

Abstract. Much time in process mining projects is spent on finding and understanding data sources and extracting the event data needed. As a result, only a fraction of time is spent actually applying techniques to discover, control and predict the business process. Moreover, current process mining techniques assume a *single* case notion. However, in real-life processes often different case notions are intertwined. For example, events of the same order handling process may refer to customers, orders, order lines, deliveries, and payments. Therefore, we propose to use *Multiple Viewpoint (MVP) models* that relate events through objects and that relate activities through classes. The required event data are much closer to existing relational databases. MVP models provide a holistic view on the process, but also allow for the extraction of classical event logs using different viewpoints. This way existing process mining techniques can be used for each viewpoint without the need for new data extractions and transformations. We provide a toolchain allowing for the discovery of MVP models (annotated with performance and frequency information) from relational databases. Moreover, we demonstrate that classical process mining techniques can be applied to any selected viewpoint.

Keywords: Process mining · Process discovery · Artifact-centric process models · Relational databases

1 Introduction

Process mining is a growing branch of data science that aims to extract insights from event data recorded in information systems. Examples of process mining techniques include process discovery algorithms that are able to find descriptive process models, conformance checking algorithms that compare event data with a given process model to find deviations, and predictive algorithms that use the discovered process model to anticipate bottlenecks or compliance problems. Gathering high-quality event data is a prerequisite for the successful application of process mining projects. However, event data are often hidden in existing information systems (e.g., the ERP systems of SAP, Microsoft, Oracle). Most systems are built on top of relational databases, to ensure data integrity

© IFIP International Federation for Information Processing 2020
Published by Springer Nature Switzerland AG 2020
P. Ceravolo et al. (Eds.): SIMPDA 2018/2019, LNBIP 379, pp. 24–51, 2020.
https://doi.org/10.1007/978-3-030-46633-6_2

and normalization. Relational databases contain entities (tables) and relations between entities. Events correspond to updates of the database, i.e., changes of the "state" of the information system. These updates may have been stored in tables of the databases (through *in-table versioning* like the change tables in SAP or any table that contains dates or timestamps) or can be retrieved using some database log (like *redo logs* [28,37] explicitly storing all database updates).

Extracting events from a database requires domain knowledge and may be very time-consuming. One of the reasons is that process mining techniques require a classical event log where each event needs to refer to a case. The case notion is used to "correlate" events and the corresponding process model shows the life-cycle model for the selected case notion. However, in the same process there may be suppliers, customers, orders, order lines, deliveries, and payments. These correspond to objects of the class model describing the database. One event may refer to a subset of such objects. A payment may refer to the payment itself, a customer, and an order. A delivery may refer to the delivery itself, but also to a subset of order lines or even multiple orders. Several views on the database could be retrieved, this means that, for the same database, several event logs and process models need to be extracted. When the process involves many different entities, the construction of the view is not easy. While there are some methods to automatically infer a/some case notion(s) from unstructured data [2,6,15], in most cases the specification happens manually. Moreover, the data extractions and transformations may be time-consuming and one quickly loses the overview. This often leads to divergence and convergence errors: events are forgotten or inadvertently duplicated leading to incorrect conclusions.

This paper introduces a new modeling technique that is able to calculate a graph where relationships between activities are shown without forcing the user to specify a case notion, since different case notions are combined in one succint diagram. The resulting models are called MVP models. Such models belong to the class of artifact-centric models [11,31] that combine data and process in a holistic manner to discover patterns and check compliance [22]. MVP models are annotated with frequency and performance information (e.g., delays) and provide a holistic view on the whole process. The colors of the relationships refer to the original classes. Using frequency based filtering and selections the MVP model can be seamlessly simplified. Any non-empty subset of classes provides a *viewpoint*. Given a viewpoint and an MVP model, we can automatically generate a classical event log and apply existing process mining techniques. Hence, the holistic view of the MVP model is complemented by detailed viewpoint models using conventional notations like Petri nets, BPMN models, process trees, or simple directly-follows graphs.

The techniques have been implemented using PM4Py (pm4py.org). In-memory computation is used to handle large data sets quickly. The approach has been evaluated using a log extracted from a real-life information system, and has proven to scale linearly with the number of events, classes and asymptotically linearly with the number of objects per class, while the execution time grows quadratically with the number of activities. This is a stark contrast with existing techniques (like OCBC models) with significantly worse complexity (Fig. 1).

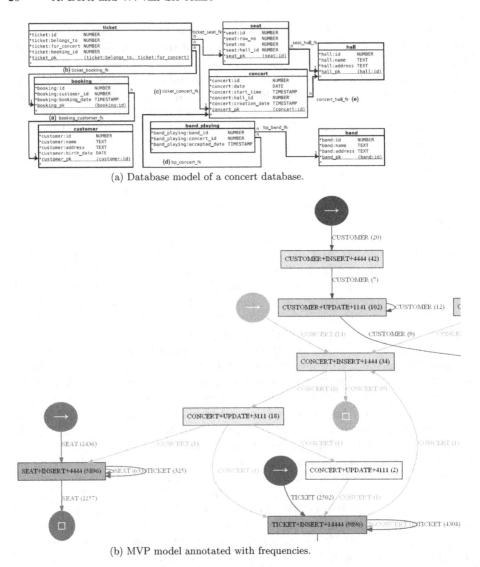

(a) Database model of a concert database.

(b) MVP model annotated with frequencies.

Fig. 1. Illustration of the approach using an example database taken from [29]. The classes are highlighted using different colors and labels on arcs. Based on a viewpoint (i.e., a set of classes) a classical event log can be generated and analyzed. (Color figure online)

In Sect. 7.4, an assessment on real-life database event logs is done. Moreover, a comparison with some existing techniques (OpenSLEX [27,29] and OCBC models [1]) is performed, considering the execution time and the usability of these approaches.

The remainder of the paper is organized as follows. Section 2 presents related work. In Sect. 3 classical and database event logs are introduced. Section 4 presents our approach to discover Multiple Viewpoint (MVP) models including Event-to-Object (E2O) graphs, Event-to-Event (E2E) multigraphs, and Activity-to-Activity (A2A) multigraphs. Section 5 introduces the notion of viewpoints and the automatic creation of classical event logs. Section 6 presents our implementation using PM4Py. In Sect. 7, we evaluate MVP models and our implementation by comparing results for real-life data sets with competing techniques.

2   Related Work

Related work may be divided into different categories:

- Approaches to extract event data from databases and to make queries easier.
- Representation of Artifact-centric models.
- Discovery of process models combining several case notions.

A few example approaches are shown in Fig. 2 and related to MVP models.

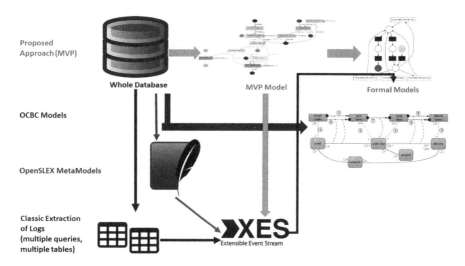

Fig. 2. Comparison of four different process mining ETL scenarios. Classical log extraction consider different queries, possibly targeting different tables, to prepare event logs for classic process mining techniques. OpenSLEX provides an easier access point to retrieve event data, but still requires the manual specification of database queries. OCBC technique extracts a single process model from the database schema, that could be useful for the purpose of understanding and checking the schema, but provide no way to retrieve a classical event log. The MVP technique, on the other hand, provides an easy visualization of the relationships between activities on top of a database, moreover it provides the possibility to get formal models (directly from the MVP models) and event logs to use with classic process mining techniques.

2.1   Related Work: Extracting Event Data from Databases

There has been earlier work on making SQL queries easier [5,8]. The basic idea
is to provide the business analyst a way to express queries in a simpler language
(SPARQL query). Some other papers related to database querying in the context
of process management are [21,25,26,33,35].

In [32], an automated technique to discover, for each notion of data object in
the process, a separate process model that describes the evolution of this object,
is presented. The technique is based on relational databases, and decomposes
the data source into multiple logs, each describing the cases of a separate data
object.

The OpenSLEX [27,29] is an high-level meta-model, that permits easier
queries, obtained from the raw database data scattered through tables (for example, the case identifier, the activity and the timestamp may be columns of different tables that need to be joined together). The aim of OpenSLEX is to let user
focus on the analysis, dealing only with elements such as events, activities, cases,
logs, objects, objects versions, object classes and attributes that are introduced
in [29]. The meta-model could be seen as a schema that captures all the pieces
of information necessary to apply process mining to database environments. To
obtain classical event logs, a case notion (connecting events to each other) needs
to be used. The OpenSLEX implementation provides indeed some connectors
for database logs (redo logs, in-table versioning, or specific database formats
[18,27]). In the implementation described in [29], OpenSLEX is supported by an
SQLite database.

2.2   Related Work: Representation of Artifact-Centric Models

Business artifacts (cf. [11,30]) combine data and process in an holistic manner as
the basic building block. These correspond to key business entities which evolve
as they pass through the business's operation.

In [4] a formal artifact-based business model, and declarative semantics based
on the use of business rules, are introduced along with a preliminary set of
technical results on the static analysis of the semantics of an artifact-based
business process model.

The Guard-Stage-Milestone (GSM) meta-model [16,17] is a formalism for
designing business artifacts in which the intended behavior is described in a
declarative way, without requiring an explicit specification of the control flow.

Some approaches which focus on compliance checking are introduced in
[19,24]. In [19], support for data-aware compliance rules is proposed in a scalable
way thanks to an abstraction approach that can serve as preprocessing step to
the actual compliance checking. In [24], compliance rule graphs are introduced,
that are a framework with support for root cause analysis, and that can provide
assistance in proactively enforcing compliance by deriving measures to render the
rule activation satisfied. In [14], conformance checking on artifact-centric processes is approached by partitioning the problem into behavioral conformance of

single artifacts and interaction conformance between artifacts, solving behavioral conformance by a reduction to existing techniques.

In [34], the concept of relational process structure is introduced, aiming to overcome some limitations of small processes such as artifacts, object lifecycles, or proclets. In these paradigms, a business process arises from the interactions between small processes. However, many-to-many relationships support is lacking. The relational process structure provides full support for many-to-many relationships and cardinality constraints at both design- and run-time.

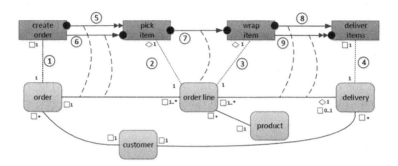

Fig. 3. Representation of a small Object-centric Behavioral Constraint (OCBC) model (taken from [1]).

2.3   Related Work: Discovery of Process Models Using Several Case Notions

In this category, we aim to describe some models that could be discovered by combining several case notions, or studying the interactions between the different case notions: interacting artifacts [23], multi-perspective models [13], object-centric models [1].

In [23], a semi-automatic approach is presented to discover the various objects supporting the system from the plain database of an ERP (Enterprise Resources Planning) system. An artifact-centric process model is identified describing the system's objects, their life-cycles, and some detailed information about interactions between objects.

Multi-instance mining [13] was introduced to discover models where different perspectives of the process can be identified. Instead of focusing on the events or activities that are executed in the context of a particular process, the focus is on the states of the different perspectives and on the discovery of the relationships between them. The *Composite State Machine Miner* [12] supports the approach [13]. It quantifies and visualises the interactions between perspectives to provide additional process insights.

Object-centric models [1] are process models, involving entities and relations in the ER model, where multiple case notions may coexist. A small OCBC model is represented in Fig. 3. In the representation, the lower part (yellow) represents the class model, and the upper part represents the activities and the constraints between them. The activities and the classes are connected by arcs if they are in relationship. OCBC models can be discovered, and can be used for compliance checking, from XOC logs. The XOC format is important because it extends the XES format with support to database-related information (related objects, relationships, state of the object model), and is one of the few choices to store database event logs along with instances of the OpenSLEX meta-model. XOC logs are using XML and contain a list of events. Each event is referring some objects, and contains the status of the database at the moment the event happened. In [20], the algorithm to infer OCBC models is described, that takes an XOC log as well as a set of possible behavioral constraint types as input, that means users can specify the constraint type set based on their needs. ProM 6 plug-ins have been realized to import XOC logs, for the discovery of a process model, and for conformance checking on top of OCBC models. The discovery algorithm can discover constraints of 9 different types.

3   Database Event Logs

Relational databases are organized in entities (classes of *objects* sharing some properties), relationships (connections between entities [10]), attributes (properties of entities and relationships). Events can be viewed as updates of a database (e.g. insertion of new objects, changes to existing objects, removal of existing objects). Some ways to retrieve events from databases are:

- Using redo logs (see [29]). These are logs where each operation in the database is saved with a timestamp; this helps to guarantee consistency, and possibility to rollback and recovery.
- Using in-table versioning. In this case, the primary key is extended with a timestamp column. For each phase of the lifecycle of an object, a new entry is added to the in-table versioning, sharing the base primary key values but with different values for the timestamp column.

An event may be linked to several objects (for example, the event that starts a marketing campaign in a CRM system may be linked to several customers), and an object may be linked to several events (for example, each customer can be related to all the tickets it opens). For the following definition, let $\mathcal{U}_E$ be the universe of events (all the events happening in a database context), $\mathcal{U}_C$ be the universe of case identifiers, $\mathcal{U}_A$ be the universe of activities (names referring to a particular step of a process), $\mathcal{U}_{attr}$ be the universe of attribute names (all the names of the attributes that can be related to an event), $\mathcal{U}_{val}$ be the universe of attribute values (all the possible values for attributes).

In this paper we consider event data closer to real-life information systems. Before providing a definition for database event logs, we define the classical event log concept.

Definition 1 (Classical Event Log). *A log is a tuple* $L = (C_I, E, A, case\_ev,$
act, attr, $\leq$) *where:*

- $C_I \subseteq \mathcal{U}_C$ *is a set of case identifiers.*
- $E \subseteq \mathcal{U}_E$ *is a set of events.*
- $A \subseteq \mathcal{U}_A$ *is the set of activities.*
- $case\_ev \in C_I \rightarrow \mathcal{P}(E) \setminus \{\emptyset\}$ *maps case identifiers onto set of events (belonging to the case).*
- $act \in E \rightarrow \mathcal{U}_A$ *maps events onto activities.*
- $attr \in E \rightarrow (\mathcal{U}_{attr} \nrightarrow \mathcal{U}_{val})$ *maps events onto a partial function assigning values to some attributes.*
- $\leq$ $\subseteq E \times E$ *defines a total order on events.*

This classical event log notion matches the XES storage format [36], that is the common source of information for process mining tools like Disco, ProcessGold, Celonis, QPR, Minit, … An example attribute of an event e is the timestamp $attr(e)(time)$ which refer to the time the event happened. While, in general, an event belongs to a single case, in Definition 1 the function $case\_ev$ might be such that cases share events.

For events extracted from a database, the function $case\_ev$ is not given, since an event may be related to different objects, and different case notions may exist. In the following Definition 2, database event logs are introduced.

Definition 2 (Database Event Log). *Let* $\mathcal{U}_O$ *be the universe of objects (all the objects that are instantiated in the database context) and* $\mathcal{U}_{OC}$ *be the universe of object classes (a class defines the structure and the behavior of a set of objects). A database event log is a tuple* $L_D = (E, O, C, A, class, act, attr, EO, \leq)$ *where:*

- $E \subseteq \mathcal{U}_E$ *is the set of events.*
- $O \subseteq \mathcal{U}_O$ *is the set of objects.*
- $C \subseteq \mathcal{U}_{OC}$ *is the set of object classes.*
- $A \subseteq \mathcal{U}_A$ *is the set of activities.*
- $class : O \rightarrow C$ *is a function that associates each object to the corresponding object class.*
- $act \in E \rightarrow A$ *maps events onto activities.*
- $attr \in E \rightarrow (\mathcal{U}_{attr} \nrightarrow \mathcal{U}_{val})$ *maps events onto a partial function assigning values to some attributes.*
- $EO \subseteq E \times O$ *relates events to sets of object references.*
- $\leq$ $\subseteq E \times E$ *defines a total order on events.*

This definition differs from Definition 1 because a case notion is missing (no $case\_ev$ function) and events are related to objects in a possible many-to-many relation. A function EO is introduced that relates events to sets of object references. Moreover, the sets of objects and classes in the event log are specified, and a function $class$ that associates each object to its class is introduced.

4   Approach to Obtain MVP Models

In this section, the different ingredients of MVP models will be introduced. The
E2O graph will be obtained directly from the database logs; the E2E multigraph will be obtained in linear complexity by calculating directly-follows relationships between events in the perspective of some object; the A2A multigraph
will be obtained in linear complexity by calculating directly-follows relationships
between activities in the perspective of some object class, using the information
stored in the E2E multigraph. The A2A multigraph is the main contributor to
the visualization of the MVP model. A projection function will be given in Sect. 5
to obtain a classical event log when a so-called viewpoint is chosen (Fig. 4).

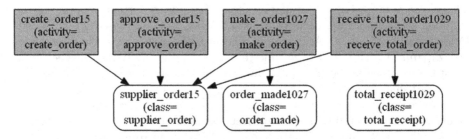

Fig. 4. Visualization of part of the E2O graph of an example database event log (found
in the erp.xoc test file). Events (red nodes) are connected to objects (white nodes).
(Color figure online)

MVP models are composed of several graphs (E2O, E2E, A2A) and auxiliary
functions (a complete definition will be presented at the end of this section),
and are constructed by reading a representation of event data retrieved from a
database (importing from intermediate structures like OpenSLEX or XOC logs).

Definition 3 (E2O Graph). *Let $L_D = (E, O, C, A, class, act, attr, EO, \leq)$ be
a database event log. The Event-to-Object graph (E2O) corresponding to the
database event log L_D can be defined as:*

$$E2O(L_D) = (E \cup O, EO)$$

*Here, the nodes are the events (E) and the objects (O), and EO (as retrieved
from the log) is a subset of $E \times O$.*

The E2O graph is obtained directly from the data without any transformation.
The remaining steps in the construction of an MVP model are the construction
of the E2E multigraph and of the A2A multigraph (Fig. 5).

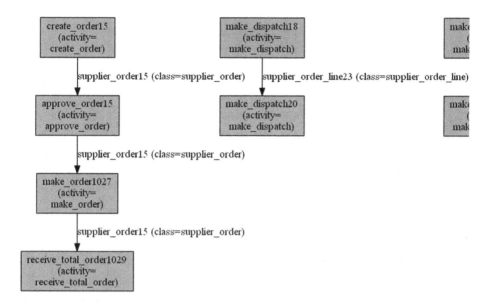

Fig. 5. Visualization of part of the E2E multigraph of an example database event log (erp.xoc test file). Events are connected to events; in the edge label, the (object) perspective has been reported.

Definition 4 (Sequence of related events). *Let* $L_D = (E, O, C, A, class,$ *act, attr, EO,* $\leq)$ *be a database event log. For* $o \in O$, *the following sequence of related events is defined:*

$$\widetilde{O}(o) = \{e_1, \ldots, e_n\}$$

such that $\{e_1, e_2, \ldots, e_n\} = \{e \mid (e, o) \in EO\}$ *and* $\forall_{1 \leq i < j \leq n} \; e_i < e_j$.

Definition 5 (E2E Multigraph). *Let* $L_D = (E, O, C, A, class, act, attr, EO,$ $\leq)$ *be a database event log. The Event-to-Event multigraph (E2E) on the database event log* L_D *can be defined as:*

$$E2E(L_D) = (E, F_E, \Pi_{perf}^E)$$

Where the nodes are the events (E) and the set of edges F_E *is defined as:*

$$F_E = \{(e_1, e_2, o) \in E \times E \times O \mid \exists_{2 \leq i \leq |\widetilde{O}(o)|} \; \widetilde{O}_{i-1}(o) = e_1, \widetilde{O}_i(o) = e_2\}$$

and $\Pi_{perf}^E : F_E \to \mathbb{R}^+ \cup \{0\}$, $\Pi_{perf}^E(e_1, e_2, o) = attr(e_2)(time) - attr(e_1)(time)$ *associates each edge to a non-negative real number expressing its duration (performance).*

The introduction of the set F_E is useful for the definition of the A2A multigraph. Although it does not make sense to represent the overall E2E multigraph, involving relationships between all events, it may be useful to display directly-follows relationships involving a small subgroup of events (Fig. 6).

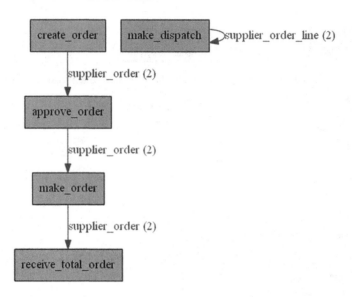

Fig. 6. Visualization of the A2A multigraph of an example database event log (erp.xoc test file). Activities are connected to activities; in the edge label, the (class) perspective along with the count of the occurrences has been reported.

Definition 6 (A2A Multigraph). *Let $L_D = (E, O, C, A, class, act, attr, EO, \leq)$ be a database event log. Let $AE : A \times A \times C \to \mathcal{P}(E \times E \times O)$ such that for $a_1, a_2 \in A$ and $c \in C$:*

$$AE(a_1, a_2, c) = \{(e_1, e_2, o) \in F_E \mid act(e_1) = a_1 \wedge act(e_2) = a_2 \wedge class(o) = c\}$$

The AE function associates to each triple (a_1, a_2, c) the set of all the events of the corresponding activities and classes. The Activity-to-Activity multigraph (A2A) on the database event log L_D can be defined as:

$$A2A(L_D) = (A, F_A, \Pi^A_{count}, \Pi^A_{perf})$$

Where the nodes are the activities (A), the set of edges F_A is defined as:

$$F_A = \{(a_1, a_2, c) \in A \times A \times C \mid AE(a_1, a_2, c) \neq \emptyset\}$$

and:

- *$\Pi^A_{count}(a_1, a_2, c) = |AE(a_1, a_2, c)|$ is the number of occurrences associated to the edge $(a_1, a_2, c) \in F_A$, that is the number of corresponding edges contained in $AE(a_1, a_2, c)$.*
- *$\Pi^A_{perf}(a_1, a_2, c) = \frac{\sum_{f_E \in AE(a_1, a_2, c)} \Pi^E_{perf}(f_E)}{\Pi^A_{count}(a_1, a_2, c)}$ is the performance associated to the edge $(a_1, a_2, c) \in F_A$, that is the average of the duration of the corresponding edges contained in $AE(a_1, a_2, c)$. An high average duration may correspond to a bottleneck in the process.*

The following definitions are useful for the representation of an MVP model, introducing clear start and end points for each class and contributing to the possibility to filter out edges.

Definition 7 (Start and End Activities of a Class). *Let* $L_D = (E, O, C, A, class, act, attr, EO, \leq)$ *be a database event log. Let* $c \in C$ *be a class. The following functions are defined:*

- $START_A(L_D) : C \to \mathcal{P}(A)$, $START_A(L_D)(c) = \{act(\widetilde{O}_1(o)) \mid o \in O \land |\widetilde{O}(o)| \geq 1 \land class(o) = c\}$ *is the set of start activities of class* c.
- $END_A(L_D) : C \to \mathcal{P}(A)$, $END_A(L_D)(c) = \{act(\widetilde{O}_{|\widetilde{O}(o)|}(o)) \mid o \in O \land |\widetilde{O}(o)| \geq 1 \land class(o) = c\}$ *is the set of end activities of class* c.

Definition 8 (Dependency Threshold between Activities given a Class). *Let* $L_D = (E, O, C, A, class, act, attr, EO, \leq)$ *be a database event log. Let* F_A *be the set of edges in* $A2A(L_D)$. *For* $(a_1, a_2, c) \in F_A$ *it is possible to define a dependency measure in the following way:*

$$dep_A(L_D) : F_A \to [0, 1]$$

$$dep_A(L_D)(a_1, a_2, c) = \begin{cases} \frac{\Pi^A_{count}(a_1, a_2, c)}{\Pi^A_{count}(a_1, a_2, c) + 1} & \text{if } a_1 = a_2 \lor (a_2, a_1, c) \notin F_A \\ \frac{\Pi^A_{count}(a_1, a_2, c) - \Pi^A_{count}(a_2, a_1, c)}{\Pi^A_{count}(a_1, a_2, c) + \Pi^A_{count}(a_2, a_1, c) + 1} & \text{if } a_1 \neq a_2 \land (a_2, a_1, c) \in F_A \end{cases}$$

Definition 9 (MVP Discovery). *Let* $L_D = (E, O, C, A, class, act, attr, EO, \leq)$ *be a database event log. We define as MVP model discovered from the log* L_D, *and we refer to it as* $MVP(L_D)$, *the following object:*

$$MVP(L_D) = (L_D, E2O(L_D), E2E(L_D), A2A(L_D), START_A(L_D),$$
$$END_A(L_D), dep_A(L_D))$$

Given a dependency threshold $d \in [-1, 1]$, a representation of an MVP model draws as many edges between a couple of activities $(a_1, a_2) \in A \times A$ as the number of classes $c \in C$ such that $dep_A(L_D)(a_1, a_2, c)$ is defined and $dep_A(L_D)(a_1, a_2, c) \geq d$.

The visualization of a MVP model is valuable (Fig. 7):

- An holistic view on the classes and the activities of the database, and on the order in which they happen, is provided.
- Arcs are decorated with frequency/performance information. Frequency information helps to understand the most frequent paths of a process, and performance information helps to discover the bottlenecks of a process.

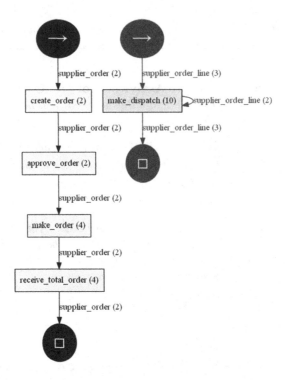

Fig. 7. Visualization of the MVP Model of an example database event log (erp.xoc test file). Activities are connected to activities; in the edge label, the (class) perspective along with the count of the occurrences has been reported. In addition to the A2A multigraph, start and end nodes are associated to each class; moreover, the edges that are reported in the A2A multigraph are filtered according to the dependency threshold (in this simple example, no arc is filtered).

5 Viewpoints: Retrieval of DFGs and Logs

The goal of this section is to provide some ways to retrieve, from an MVP model $MVP(L_D)$, built upon the database event log L_D, a particular viewpoint on the model.

Definition 10 (Viewpoint). *Let* $L_D = (E, O, C, A, class, act, attr, EO, \leq)$ *be a database event log. Let*

$$MVP(L_D) = (L_D, E2O(L_D), E2E(L_D), A2A(L_D), START_A(L_D),$$
$$END_A(L_D), dep_A(L_D))$$

be an MVP model. A viewpoint is a set of classes $V(L_D) \subseteq C$. *A viewpoint is corresponding to a subset of edges in the E2E graph (that is* F_E*):*

$$V_E(L_D) = \{(e_1, e_2, o) \in F_E \mid o \in V(L_D)\}$$

From a view, we can obtain two different final outputs.

– A Directly-Follows Graph (DFG), that includes the edges related to the classes contained in $V(L_D)$.
– A classical event log, that includes all the events that are related to the objects having a class contained in $V(L_D)$.

An example of projection could be found in Fig. 8, where a singleton viewpoint containing a single class is chosen, a Directly-Follows Graph is obtained and a Petri net is obtained through the application of Inductive Miner Directly-Follows.

5.1   Projection on a Directly-Follows Graph

The concept of a Directly-Follows Graph is introduced in the following definition:

Definition 11 (DFG). *A Directly-Follows Graph is a weighted directed graph:*

$$DFG = (N_{DFG}, E_{DFG}, c_{DFG})$$

Where N_{DFG} (the nodes) are the activities, and $E_{DFG} \subseteq N_{DFG} \times N_{DFG}$ is the set of all the edges between activities that happened in direct succession, and $c_{DFG} : E_{DFG} \to \mathbb{R}^+$ is the count function that aims to represent how many times two different activities happened in direct succession.

Fig. 8. Petri net extracted using Inductive Miner Directly-Follows from the projection of the MVP model extracted from the erp.xoc test file on the viewpoint containing the *supplier_order* class.

From a viewpoint $V(L_D)$ on MVP(L_D), it is possible to obtain a Directly-Follows Graph as explained in the following definition:

Definition 12 (DFG given a viewpoint). *Given a viewpoint $V(L_D)$ on MVP(L_D), a DFG could be obtained taking:*

– $N_{DFG} = \bigcup_{(e_1, e_2, o) \in V_E(L_D)} \{act(e_1), act(e_2)\}$
– $E_{DFG} = \bigcup_{(e_1, e_2, o) \in V_E(L_D)} \{(act(e_1), act(e_2))\}$
– $c_{DFG}(a_1, a_2) = |\{(e_1, e_2, o) \in V_E(L_D) \mid act(e_1) = a_1 \wedge act(e_2) = a_2\}|$

where $V_E(L_D)$ is obtained as in Definition 10.

An example could be provided. Let $V_E(L_D) = \{(e_1, e_2, o_1), (e_2, e_3, o_1), (e_4,$ $e_5, o_2), (e_5, e_6, o_2)\}$ be the set of edges associated to a viewpoint, such that:

- $\text{act}(e_1) = \text{act}(e_4) = A$
- $\text{act}(e_2) = \text{act}(e_5) = B$
- $\text{act}(e_3) = C$
- $\text{act}(e_6) = D$

Then the DFG is such that $N_{\text{DFG}} = \{A, B, C, D\}$, $E_{\text{DFG}} = \{(A, B), (B, C),$ $(B, D)\}$, $c_{\text{DFG}}(B, C) = c_{\text{DFG}}(B, D) = 1$, $c_{\text{DFG}}(A, B) = 2$.

5.2   Projection on a Log

The projection of an MVP model $\text{MVP}(L_D)$ obtained from a database event log L_D to a classical event log (see Sect. 3) could be introduced when a viewpoint $V(L_D) \subseteq C$ is chosen.

Indeed, the information contained in an MVP model could be used to determine a case notion C_D that is used to transform the database event log L_D into a classical event log, in a way that events belonging to the same process execution can be grouped.

The definition of case notion on database event logs could be introduced:

Definition 13 (Case Notion). *Let $L_D = (E, O, C, A, class, act, attr, EO, \leq)$ be a database event log. A case notion is a set of sets of events $C_D \subseteq \mathcal{P}(E) \setminus \{\emptyset\}$.*

The case notion does not need to cover all the events contained in E, moreover the intersection between sets of events contained in the case notion may also not be empty.

An example of case notion could be provided. Let $E = \{e_1, e_2, e_3, e_4, e_5, e_6\}$ be a set of events. Then a case notion might be $C_D = \{\{e_1, e_2\}, \{e_3, e_4\}, \{e_1, e_4, e_5\}\}$. Let's note that the union of all these sets is not E, and the intersection between $\{e_1, e_2\}$ and $\{e_1, e_4, e_5\}$ is not empty.

When a case notion is defined, it is possible to define the projection function from the database event log L_D to the classical event log. An assumption is that the case notion C_D is contained in the universe of case identifiers $\mathcal{U}_C$. This helps to define the function case_ev in a simpler way.

Definition 14 (Projection function). *Let $L_D=(E, O, C, A, class, act, attr, EO, \leq)$ be an event log in a database context. Let $C_D \subseteq \mathcal{P}(E) \setminus \{\emptyset\}$ be a case notion. Then it is possible to define a projection function from a database event log to a classical event log as:*

$$proj(L_D, C_D) = (C_D, E, A, case\_ev, act, attr, \leq)$$

where $case\_ev \in C_D \rightarrow \mathcal{P}(E) \setminus \{\emptyset\}$ is such that for all $c \in C_D$, $case\_ev(c) = c$.

Given an MVP model $MVP(L_D)$ and a viewpoint $V(L_D)$ on that defines a set of edges $V_E(L_D)$ in the E2E multigraph, a case notion is defined as:

$$C_D = \left\{ \bigcup_{o' \in O, \text{class}(o') \in V(L_D), \widetilde{O}(o) \cap \widetilde{O}(o') \neq \emptyset} \widetilde{O}(o') \mid o \in O, \text{class}(o) \in V(L_D) \right\}$$

A classical event log is obtained as $L = \text{proj}(L_D, C_D)$.

6   Tool

MVP Models have been implemented in a feature branch of the PM4Py Process Mining library[1] [3]. The architecture of the tool provides a clear separation between the management of the log (object), the MVP discovery algorithm and the MVP visualization. Moreover, utilities have been provided to generate a database event log and to visualize the E2O and the E2E multigraphs. A reference technical manual with a description of the features provided in tool is contained in http://www.alessandroberti.it/technical.pdf.

The provided features are:

- **Log management:** log importing (XOC, OpenSLEX, Parquet), log exporting (XOC, Parquet).
- **Model discovery:** discovery of a MVP model with frequency or performance decoration.
- **Visualization:** E2O graphs, E2E and A2A multigraphs, MVP models (with possibility to filter out edges with a dependency measure that is below the threshold).
- **Database log generation:** the option to generate an MVP model specifying the number of events, activities, classes and a number of objects for each class is provided.
- **Projection on a Viewpoint:** projection on a DFG and on a log.
- **Storage of MVP models:** importing/exporting of MVP models into a dump file.

Example logs are provided in the tests folder of the repository, in particular:

- *logOpportunities.parquet* is the database log used in the assessment and extracted from the Dynamics CRM system.
- *metamodel.slexmm* provides the OpenSLEX metamodel of the Concert database.
- *erp.xoc* provides an example XOC log extracted from a Dollibar ERP system.

The storage used in the tool for the database event logs is represented in Fig. 9. Several rows are associated to an event, and contain an ID, an activity, a timestamp, and a single object identifier in the column corresponding to its class. This permits to store events in a tabular format using basic types for the columns (strings, integers, dates) to maximize the query performance.

---

[1] The repository can be accessed at the URL https://github.com/Javert899/pm4py-source.

event_id	event_activity	event_timestamp	supplier_order	supplier_order_line
create_order15	create_order	2016-10-21 11:38:26	supplier_order15	NaN
approve_order15	approve_order	2016-10-21 11:38:53	supplier_order15	NaN
make_order1027	make_order	2016-10-21 11:40:00	NaN	NaN
make_order1027	make_order	2016-10-21 11:40:00	supplier_order15	NaN
make_dispatch19	make_dispatch	2016-10-21 11:42:31	NaN	NaN
make_dispatch19	make_dispatch	2016-10-21 11:42:31	NaN	supplier_order_line22
make_dispatch18	make_dispatch	2016-10-21 11:42:31	NaN	NaN
make_dispatch18	make_dispatch	2016-10-21 11:42:31	NaN	supplier_order_line23
receive_partial_order1028	receive_partial_order	2016-10-21 11:43:00	NaN	NaN
make_dispatch21	make_dispatch	2016-10-21 11:44:25	NaN	NaN
make_dispatch21	make_dispatch	2016-10-21 11:44:25	NaN	supplier_order_line22
make_dispatch20	make_dispatch	2016-10-21 11:44:25	NaN	NaN
make_dispatch20	make_dispatch	2016-10-21 11:44:25	NaN	supplier_order_line23
receive_total_order1029	receive_total_order	2016-10-21 11:45:00	NaN	NaN
receive_total_order1029	receive_total_order	2016-10-21 11:45:00	supplier_order15	NaN
create_invoice17	create_invoice	2016-10-21 11:45:46	NaN	NaN
create_invoice17	create_invoice	2016-10-21 11:45:46	NaN	NaN
create_payment11	create_payment	2016-10-21 11:46:14	NaN	NaN
create_payment11	create_payment	2016-10-21 11:46:14	NaN	NaN
create_payment12	create_payment	2016-10-21 11:46:29	NaN	NaN
create_payment12	create_payment	2016-10-21 11:46:29	NaN	NaN
create_order16	create_order	2016-10-21 11:56:35	supplier_order16	NaN
approve_order16	approve_order	2016-10-21 11:56:50	supplier_order16	NaN
make_order1033	make_order	2016-10-21 11:57:00	NaN	NaN
make_order1033	make_order	2016-10-21 11:57:00	supplier_order16	NaN
make_dispatch22	make_dispatch	2016-10-21 11:57:28	NaN	NaN
make_dispatch22	make_dispatch	2016-10-21 11:57:28	NaN	supplier_order_line24
receive_total_order1034	receive_total_order	2016-10-21 11:58:00	NaN	NaN
receive_total_order1034	receive_total_order	2016-10-21 11:58:00	supplier_order16	NaN
create_invoice18	create_invoice	2016-10-21 11:58:36	NaN	NaN
create_invoice18	create_invoice	2016-10-21 11:58:36	NaN	NaN

Fig. 9. Representation of a database log in the Parquet columnar format. This is different from classic CSV event logs since the same event is repeated in multiple rows, one for each related object. Both information related to events and related objects are columns of a table. This structure reflects the way information is stored inside the in-memory Pandas dataframe that supports the filtering and the discovery operations.

7   Assessment

The assessment of MVP models will show how they perform against two competing approaches (OCBC models [20] and OpenSLEX [29]), both considering the execution time and the simplicity/readability/amount of information contained in the models. These approaches have been chosen because are very recent. The scalability of the MVP models implementation contained in PM4Py will be analyzed on some synthetic logs. Moreover, an assessment using a real information system (Microsoft Dynamics CRM) will be shown.

7.1   Comparison with Related Approaches

OCBC models are powerful descriptions of the relationships between activities and classes at the database level. Having said that, OCBC models have scalability issues: XOC logs as proposed in [1] store a snapshot of the object model per event, this becomes quickly unfeasible also for few hundred database events (as it will shown in the assessment). In [9], an updated version of XOC, storing in each event the updates to the object model, has been proposed, although a XOC event log in such format is not public available. Moreover, the final visualization (even if it can be filtered on some types of constraints) lacks understandability. A serious issue is the lack of support for frequency/performance decoration (i.e. the number of occurrences of the arc, the time passed between activity).

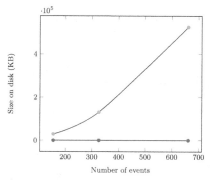

a) Difference in storage required for storing MVP logs (the line with green dots) and XOC logs (the line with orange dots) for different data sets.

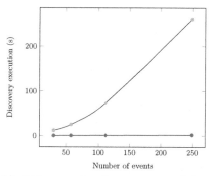

b) Difference in discovery speed between MVP models (the line with green dots) and OCBC models (the line with orange dots) for different data sets.

Fig. 10. Comparison between MVP models and OCBC models, in (a) size on disk of the proposed log storage (b) discovery speed. The XOC format that is tested is the one described in [1]. (Color figure online)

OpenSLEX provides a way to ingest a database event log into a meta-model instance that is easier to query. An issue is that they do not offer any visual clue on underlying relationships between activities, making the life difficult to the user in first instance. Moreover, although the queries on OpenSLEX are easier than the queries done directly on the database, there is an effort required by the user to understand the concepts described in [29].

7.2   Evaluation of the Execution Time

OCBC models (tests have been performed on the original version of XOC, proposed in [1], for which several logs are public available) have scalability problems with regards to the log format (XOC), that requires the storage, for each event, of the status of the entire object model. In Fig. 10(a), the size on disk (in KB) of a log containing the specified number of events has been considered. To obtain an OCBC model, for a log with just 661 events, a XOC log of 512 MB is required, that is 17476 greater than the amount of disk space that permits to discover an MVP model from the same database.

In Fig. 10(b), the execution speed of the discovery procedure has been compared between MVP models and OCBC models. To obtain an OCBC model, for a log with just 249 events, a time of 4 min and 20 s is required, while for discovering an MVP model from the same database 2 s are required, that is 127 times faster. This result holds also for the new version of the XOC format [9], since in the later version the object model is built in-memory by the discovery algorithm starting from the updates described in the event log.

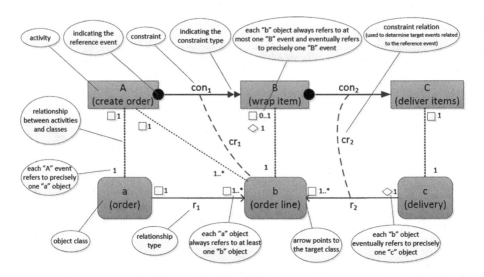

Fig. 11. Usability assessment: example model illustrating the main ingredients of OCBC models.

With OpenSLEX, the comparison is more tight, although the final goal is different: OpenSLEX require a query to get a classical event log, while MVP models do not require this effort. There is only a meta-model instance available in public, that has been extracted from a synthetic database on concert management. The amount of storage required to store the OpenSLEX instance is 11 MB, while the amount of storage that permits to discover an MVP model is 222 KB (storing using Parquet format). To compare the (time) performance of OpenSLEX and MVP models, a query on the OpenSLEX instance needs to be performed. Starting in both cases from the concert database instance of the OpenSLEX meta-model, if an example query[2] is chosen, the execution speed of the query on the OpenSLEX instance is 1.73 s. The time needed for MVP models to obtain a complete model with frequency and performance information is 1.96 s, this without requiring the specification of any query by the user.

7.3 Usability of the Approaches

OCBC and MVP models both provide ways to discover a model on top of database event logs. The amount of information and constraints extracted by OCBC models is very high, and although filtering is provided to keep only some constraints, the resulting process model is complex to understand. An explanation of some ingredients of OCBC models is provided by [20] and represented in Fig. 11. This amount of information is insane, but it is very difficult also for a process analyst to be able to understand it without proper training.

---

[2] The example query is provided in the technical report available at the address http://www.alessandroberti.it/technical.pdf.

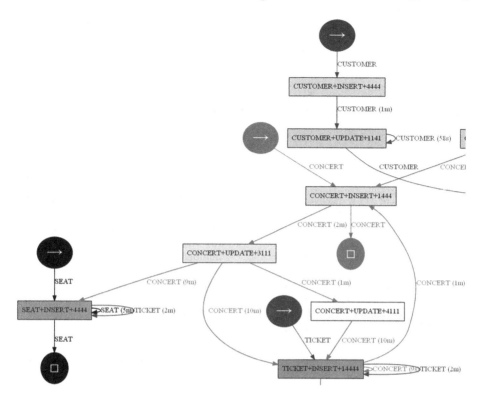

Fig. 12. Usability assessment: performance/bottleneck visualization provided by MVP models.

Moreover, this class of models does not provide frequency/performance information, that can be useful for the process analyst in order to detect the bottlenecks. MVP models, as represented in Fig. 12, can provide a graph in which edges are decorated by frequency/performance information. With MVP models, the following features are also available, that are not available on OCBC models:

– Projection to a classic directly-follows graph to be used with techniques like Inductive Miner Directly-Follows and the Heuristics Miner.
– Projection to a classical event log to be used with mainstream process mining techniques.

OpenSLEX do not provide a visualization of a process model on top of the database, but require to the user the specification of a query to retrieve an event log. This is an unavoidable step, and requires time and expertise by the user. So, the retrieval of a process model using MVP requires less time and less knowledge than the retrieval from OpenSLEX.

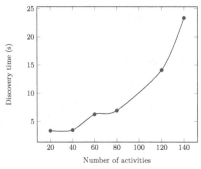

a) Performance of MVP discovery with the increase of the number of activities, when the number of classes, the number of objects per class and the number of events in the log is kept fixed: n_events = 100000, n_classes = 10 and n_objects_per_class = 1000. The logs have been obtained through the generator included in PM4Py. The execution time (in seconds) grows quadratically with the number of activities.

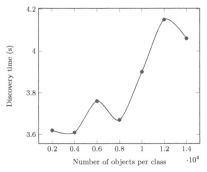

b) Performance of MVP discovery with the increase of the number of objects per class, when the number of classes, the number of activities and the number of events in the log is kept fixed: n_events = 100000, n_classes = 10 and n_activities = 40. The logs have been obtained through the generator included in PM4Py.

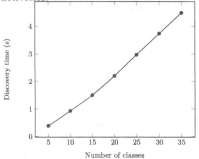

c) Performance of MVP discovery with the increase of the number of classes, when the number of objects per class, the number of activities and the number of events in the log is kept fixed: n_events = 10000, n_activities = 40 and n_objects_per_class = 6000. The logs have been obtained through the generator included in PM4Py. The execution time (in seconds) grows linearly with the number of classes.

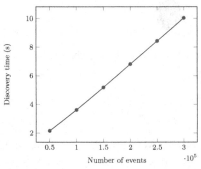

d) Performance of MVP discovery with the increase of the number of events in the log, when the number of classes, the number of objects per class and the number of activities in the log is kept fixed: n_activities = 40, n_classes = 10 and n_objects_per_class = 6000. The logs have been obtained through the generator included in PM4Py. The execution time (in seconds) grows linearly with the number of events.

Fig. 13. Scalability assessment of MVP models, with regards to (a) the number of activities (b) the number of objects per class (c) the number of classes (d) the number of events in the log.

7.4   Scalability of the Approach

In the previous section, MVP models have shown greater scalability and usability in comparison to some competing approaches. In this section, the goal is to understand more clearly the performance of the current implementation.

An assessment of the approach on simulated logs (through the log generator included in PM4Py) has been done (see Fig. 13) to see which variables influence the execution time in a quadratic way, and which variables influence it in a linear way:

- (a) assesses the performance of MVP discovery with the increase of the number of activities, when the number of classes, the number of objects per class and the number of events in the log is kept fixed. The execution time (in seconds) grows quadratically with the number of activities.
- (b) assesses the performance of MVP discovery with the increase of the number of objects per class, when the number of classes, the number of activities and the number of events in the log is kept fixed. Although the behavior in the figure looks erratic, the execution time (in seconds) grows asymptotically linearly with the number of objects per class.
- (c) assesses the performance of MVP discovery with the increase of the number of classes, when the number of objects per class, the number of activities and the number of events in the log is kept fixed. The execution time (in seconds) grows linearly with the number of classes.
- (d) assesses the performance of MVP discovery with the increase of the number of events in the log, when the number of classes, the number of objects per class and the number of activities in the log is kept fixed. The execution time (in seconds) grows linearly with the number of events.

7.5   Evaluation Using CRM Data

This section presents a study of data extracted from a Microsoft Dynamics CRM demo and analyzed using MVP discovery. A Customer Relationship Management system (CRM) [7] is an information system used to manage the commercial lifecycle of an organization, including management of customers, opportunities and marketing campaigns.

Many companies actually involve a CRM system for helping business and sales people to coordinate, share information and goals. Data extracted from Microsoft Dynamics CRM is particularly interesting since this product manages several processes of the business side, providing the possibility to define workflows and to measure KPI also through connection to the Microsoft Power BI business intelligence tool. For evaluation purposes, a database log has been generated containing data extracted from a Dynamics CRM demo. The database supporting the system contains several entities, and each entity contains several entries related to activities happening in the CRM. Each entry could be described by a unique identifier (UUID), the timestamp of creation/change, the resource that created/modified the entry, and some UUIDs of other entries belonging to

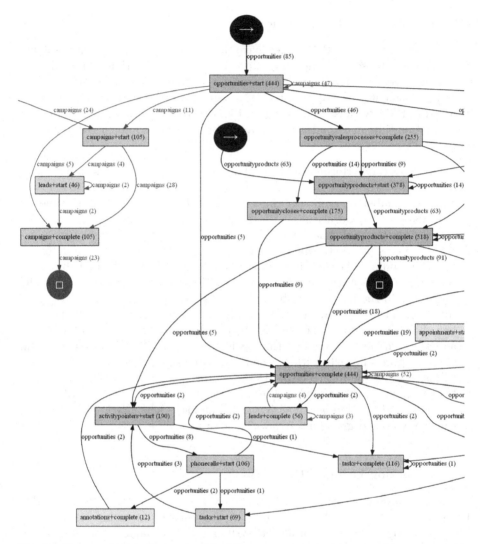

Fig. 14. Representation in one model of four perspectives (opportunities, opportunityproducts, campaigns, connections) of the Dynamics CRM database (only part of the diagram is reported).

the same or to different entities. Moreover, each entry is uniquely associated with the entity it belongs to.

The following strategy has been pursued in order to generate a log:

- For each entry belonging to an entity, two events have been associated: creation event (with the timestamp of creation and lifecycle *start*) and modify event (with the timestamp of modification and lifecycle *complete*).
- Each entry belonging to an entity has also been associated with an object.

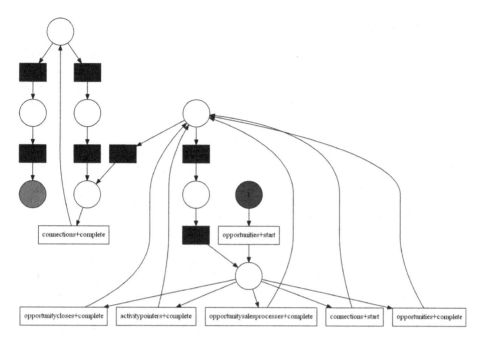

Fig. 15. Petri net obtained by applying Inductive Miner Directly-Follows on the projection of the Dynamics CRM MVP model on a viewpoint containing only the 'opportunities' class. Black boxes represent invisible transitions (transitions that can be executed without correspondence with the activities of the log).

– Relationships between events and objects are created accordingly to the relationships expressed by the entries (an entry may cite several UUIDs of other entries stored in the database).

The previous construction means that for the same entry there are two events (start+complete) and one object in the log.

The database log contains 5863 events, 4413 objects, 120 activities and 80 object classes, and could be stored in a 386 KB Parquet file. The complete MVP model (525 edges) can be calculated and represented in 5 s. Taking a viewpoint containing only classes related to opportunities management (e.g. *opportunities, opportunityproducts, campaigns, connections*), and choosing the frequency metric, the model obtained is represented in Fig. 14. Projecting the MVP model on a viewpoint containing only the 'opportunities' class and applying Inductive Miner Directly-Follows, the process model represented in Fig. 15 is obtained.

Being able to handle this complex database schema and visualize a model that unifies the different classes, in a very reasonable time, is a thing that is impossible with competing approaches like OCBC models (due to severe scalability issues) and OpenSLEX instances (because the resulting query would be more complex than selecting a viewpoint from an holistic model).

8   Conclusion

This paper introduces Multiple Viewpoint models (MVP), providing an holistic view on a process supported by a database. MVP models are annotated with frequency and performance measures (e.g., delays), supporting the detection of the most frequent paths and of the bottlenecks without the specification of any case notion. At the same time, a viewpoint (a non-empty subset of classes) can be chosen on the MVP models in order to get classical objects (DFGs and logs) to use with the mainstream process mining techniques. Hence, the holistic view of the MVP model is complemented by detailed viewpoint models using conventional notations like Petri nets, BPMN models and process trees. This possibility is not provided by the competing techniques described in Sect. 2: the focus is either on the specification of a case notion, or on the retrieval of an artifact-centric model. However, the relationships between the classes are not calculated and represented in the process model, in contrast to the technique of OCBC models. Moreover, the lack of a clear execution semantic of MVP models makes the application of conformance checking techniques possible only after choosing a viewpoint, in contrast with the OCBC technique that provides a (theoretically) powerful conformance checking approach on top of the model.

An MVP model has been discovered from a database event log extracted from a Microsoft Dynamics Customer Relationship Management (CRM) system, showing the possibility to apply the techniques described in this paper to real-life information systems.

Moreover, a comparison considering execution time and usability has been performed against two of the most recent process mining approaches on databases (OpenSLEX and OCBC models); MVP models are relatively well performing and easy to use, since an holistic view could be obtained without any effort from the user, and any viewpoint could be chosen on the MVP model.

Scalability testing proved that MVP models scale linearly with the number of events, classes and asymptotically linearly with the number of objects contained in the database event log, while they scale quadratically with the number of activities (due to the edges calculation in the A2A multigraph). This is a remarkable result in comparison to OCBC models that show an exponential complexity on the number of events (they become unmanageable also for a small number of events).

The techniques described in this paper could in principle be implemented starting from the logs of any relational database. Hence MVP discovery supports process mining analysis directly from real-life complex information systems.

References

1. van der Aalst, W., Li, G., Montali, M.: Object-centric behavioral constraints. arXiv preprint arXiv:1703.05740 (2017)
2. Bayomie, D., Helal, I.M.A., Awad, A., Ezat, E., ElBastawissi, A.: Deducing case IDs for unlabeled event logs. In: Reichert, M., Reijers, H.A. (eds.) BPM 2015. LNBIP, vol. 256, pp. 242–254. Springer, Cham (2016). https://doi.org/10.1007/978-3-319-42887-1_20

3. Berti, A., van Zelst, S.J., van der Aalst, W.: Process Mining for Python (PM4Py): bridging the gap between process- and data science 13–16 (2019)

4. Bhattacharya, K., Gerede, C., Hull, R., Liu, R., Su, J.: Towards formal analysis of artifact-centric business process models. In: Alonso, G., Dadam, P., Rosemann, M. (eds.) BPM 2007. LNCS, vol. 4714, pp. 288–304. Springer, Heidelberg (2007). https://doi.org/10.1007/978-3-540-75183-0_21

5. Bouchou, B., Niang, C.: Semantic mediator querying. In: Proceedings of the 18th International Database Engineering and Applications Symposium, pp. 29–38. ACM (2014)

6. Burattin, A., Vigo, R.: A framework for semi-automated process instance discovery from decorative attributes. In: 2011 IEEE Symposium on Computational Intelligence and Data Mining (CIDM), pp. 176–183. IEEE (2011)

7. Buttle, F.: Customer Relationship Management. Routledge, Abingdon (2004)

8. Calvanese, D., et al.: Ontop: answering SPARQL queries over relational databases. Semant. Web **8**(3), 471–487 (2017)

9. Li, G., de Murillas, E.G.L., de Carvalho, R.M., van der Aalst, W.M.P.: Extracting object-centric event logs to support process mining on databases. In: Mendling, J., Mouratidis, H. (eds.) CAiSE 2018. LNBIP, vol. 317, pp. 182–199. Springer, Cham (2018). https://doi.org/10.1007/978-3-319-92901-9_16

10. Chen, P.P.S.: The entity-relationship model toward a unified view of data. ACM Trans. Database Syst. (TODS) **1**(1), 9–36 (1976)

11. Cohn, D., Hull, R.: Business artifacts: a data-centric approach to modeling business operations and processes. IEEE Data Eng. Bull. **32**(3), 3–9 (2009)

12. van Eck, M.L., Sidorova, N., van der Aalst, W.: Composite state machine miner: discovering and exploring multi-perspective processes. In: BPM (Demos), pp. 73–77 (2016)

13. van Eck, M.L., Sidorova, N., van der Aalst, W.M.P.: Discovering and exploring state-based models for multi-perspective processes. In: La Rosa, M., Loos, P., Pastor, O. (eds.) BPM 2016. LNCS, vol. 9850, pp. 142–157. Springer, Cham (2016). https://doi.org/10.1007/978-3-319-45348-4_9

14. Fahland, D., de Leoni, M., van Dongen, B.F., van der Aalst, W.M.P.: Behavioral conformance of artifact-centric process models. In: Abramowicz, W. (ed.) BIS 2011. LNBIP, vol. 87, pp. 37–49. Springer, Heidelberg (2011). https://doi.org/10.1007/978-3-642-21863-7_4

15. Helal, I.M., Awad, A., El Bastawissi, A.: Runtime deduction of case ID for unlabeled business process execution events. In: 2015 IEEE/ACS 12th International Conference of Computer Systems and Applications (AICCSA), pp. 1–8. IEEE (2015)

16. Hull, R., et al.: Business artifacts with guard-stage-milestone lifecycles: managing artifact interactions with conditions and events. In: Proceedings of the 5th ACM International Conference on Distributed Event-Based System, pp. 51–62. ACM (2011)

17. Hull, R., et al.: Introducing the guard-stage-milestone approach for specifying business entity lifecycles. In: Bravetti, M., Bultan, T. (eds.) WS-FM 2010. LNCS, vol. 6551, pp. 1–24. Springer, Heidelberg (2011). https://doi.org/10.1007/978-3-642-19589-1_1

18. Ingvaldsen, J.E., Gulla, J.A.: Preprocessing support for large scale process mining of SAP transactions. In: ter Hofstede, A., Benatallah, B., Paik, H.-Y. (eds.) BPM 2007. LNCS, vol. 4928, pp. 30–41. Springer, Heidelberg (2008). https://doi.org/10.1007/978-3-540-78238-4_5

19. Knuplesch, D., Ly, L.T., Rinderle-Ma, S., Pfeifer, H., Dadam, P.: On enabling data-aware compliance checking of business process models. In: Parsons, J., Saeki, M., Shoval, P., Woo, C., Wand, Y. (eds.) ER 2010. LNCS, vol. 6412, pp. 332–346. Springer, Heidelberg (2010). https://doi.org/10.1007/978-3-642-16373-9_24

20. Li, G., de Carvalho, R.M., van der Aalst, W.M.P.: Automatic discovery of object-centric behavioral constraint models. In: Abramowicz, W. (ed.) BIS 2017. LNBIP, vol. 288, pp. 43–58. Springer, Cham (2017). https://doi.org/10.1007/978-3-319-59336-4_4

21. Liu, D., Pedrinaci, C., Domingue, J.: Semantic enabled complex event language for business process monitoring. In: Proceedings of the 4th International Workshop on Semantic Business Process Management, pp. 31–34. ACM (2009)

22. Lohmann, N.: Compliance by design for artifact-centric business processes. In: Rinderle-Ma, S., Toumani, F., Wolf, K. (eds.) BPM 2011. LNCS, vol. 6896, pp. 99–115. Springer, Heidelberg (2011). https://doi.org/10.1007/978-3-642-23059-2_11

23. Lu, X., Nagelkerke, M., van de Wiel, D., Fahland, D.: Discovering interacting artifacts from ERP systems (extended version). BPM reports 1508 (2015)

24. Ly, L.T., Rinderle-Ma, S., Knuplesch, D., Dadam, P.: Monitoring business process compliance using compliance rule graphs. In: Meersman, R., et al. (eds.) OTM 2011. LNCS, vol. 7044, pp. 82–99. Springer, Heidelberg (2011). https://doi.org/10.1007/978-3-642-25109-2_7

25. Metzke, T., Rogge-Solti, A., Baumgrass, A., Mendling, J., Weske, M.: Enabling semantic complex event processing in the domain of logistics. In: Lomuscio, A.R., Nepal, S., Patrizi, F., Benatallah, B., Brandić, I. (eds.) ICSOC 2013. LNCS, vol. 8377, pp. 419–431. Springer, Cham (2014). https://doi.org/10.1007/978-3-319-06859-6_37

26. Momotko, M., Subieta, K.: Process query language: a way to make workflow processes more flexible. In: Benczúr, A., Demetrovics, J., Gottlob, G. (eds.) ADBIS 2004. LNCS, vol. 3255, pp. 306–321. Springer, Heidelberg (2004). https://doi.org/10.1007/978-3-540-30204-9_21

27. de Murillas, E.G.L.: Process mining on databases: extracting event data from real life data sources (2019)

28. González López de Murillas, E., Hoogendoorn, G.E., Reijers, H.A.: Redo log process mining in real life: data challenges & opportunities. In: Teniente, E., Weidlich, M. (eds.) BPM 2017. LNBIP, vol. 308, pp. 573–587. Springer, Cham (2018). https://doi.org/10.1007/978-3-319-74030-0_45

29. González López de Murillas, E., Reijers, H.A., van der Aalst, W.M.P.: Connecting databases with process mining: a meta model and toolset. Softw. Syst. Model. 18(2), 1209–1247 (2018). https://doi.org/10.1007/s10270-018-0664-7

30. Narendra, N.C., Badr, Y., Thiran, P., Maamar, Z.: Towards a unified approach for business process modeling using context-based artifacts and web services. In: 2009 IEEE International Conference on Services Computing, pp. 332–339. IEEE (2009)

31. Nigam, A., Caswell, N.S.: Business artifacts: an approach to operational specification. IBM Syst. J. 42(3), 428–445 (2003)

32. Nooijen, E.H.J., van Dongen, B.F., Fahland, D.: Automatic discovery of data-centric and artifact-centric processes. In: La Rosa, M., Soffer, P. (eds.) BPM 2012. LNBIP, vol. 132, pp. 316–327. Springer, Heidelberg (2013). https://doi.org/10.1007/978-3-642-36285-9_36

33. Song, L., Wang, J., Wen, L., Wang, W., Tan, S., Kong, H.: Querying process models based on the temporal relations between tasks. In: 2011 IEEE 15th International Enterprise Distributed Object Computing Conference Workshops, pp. 213–222. IEEE (2011)

34. Steinau, S., Andrews, K., Reichert, M.: The relational process structure. In: Krogstie, J., Reijers, H.A. (eds.) CAiSE 2018. LNCS, vol. 10816, pp. 53–67. Springer, Cham (2018). https://doi.org/10.1007/978-3-319-91563-0_4
35. Tang, Y., Mackey, I., Su, J.: Querying workflow logs. Information **9**(2), 25 (2018)
36. Verbeek, H.M.W., Buijs, J.C.A.M., van Dongen, B.F., van der Aalst, W.M.P.: XES, XESame, and ProM 6. In: Soffer, P., Proper, E. (eds.) CAiSE Forum 2010. LNBIP, vol. 72, pp. 60–75. Springer, Heidelberg (2011). https://doi.org/10.1007/978-3-642-17722-4_5
37. Yano, K., Nomura, Y., Kanai, T.: A practical approach to automated business process discovery. In: 2013 17th IEEE International Enterprise Distributed Object Computing Conference Workshops, pp. 53–62. IEEE (2013)

Standardizing Process-Data Exploitation by Means of a Process-Instance Metamodel

Antonio Cancela$^{(\boxtimes)}$ ⓘ, Antonia M. Reina Quintero ⓘ,
María Teresa Gómez-López ⓘ, and Alejandro García-García

Universidad de Sevilla, Sevilla, Spain
{acancela,reinaqu,maytegomez}@us.es, algarcia@emergya.com
http://www.idea.us.es

Abstract. The analysis of data produced by enterprises during business-process executions is crucial in ascertaining how these processes work and how they can be optimized, despite heterogeneous nature of these data structures. This data may also be used for various types of analysis, such as reasoning, process querying and process mining, which consume different data formats. However, all these structures and formats share a common ground: the business-process model and its instantiation are in each of their kernels. In this paper, we propose the use of a Business-Process Instance Metamodel, which serves as a common interface to perform an independent exploitation of data from the applications that produce the data and those which consume the data. A tool has been implemented as a proof of concept to illustrate the ease of matching the data with the proposed metamodel.

Keywords: Process-Instance Metamodel · Data model · Model mapping · Domain knowledge · Business-process data exploitation

1 Introduction

Companies today produce a great amount of data in a daily basis that accurately reflects their business processes. This data holds special interest for the study and optimization of these business processes. However, increasingly, the data comes from heterogeneous sources with different formats and structures (relational databases, NoSQL, APIs, data warehouses...). This causes the analysis and exploitation of data to become highly time-consuming, since many solutions are ad-hoc solutions, and, as a consequence, they have to be adapted depending on the techniques to be applied. This data-preparation process constitutes the most significant barrier to be improved and one of the highest time-consuming tasks in data-analysis projects [38].

In this paper, we propose the use of a Business-Process Instance Metamodel as an intermediate layer to specify the relation between the domain-specific data

© IFIP International Federation for Information Processing 2020
Published by Springer Nature Switzerland AG 2020
P. Ceravolo et al. (Eds.): SIMPDA 2018/2019, LNBIP 379, pp. 52–66, 2020.
https://doi.org/10.1007/978-3-030-46633-6_3

produced and its meaning in a business process, thereby facilitating how it can be exploited by business analysis techniques. Our research goal is *the simplification of the data analysis by making independent the structures of data production from data consumption*. The approach is based on the definition of mappings between data sources and the business process concepts specified in the Business-Process Instance Metamodel. The benefits obtained by using an intermediate metamodel include the reduction of the analysis time and the exploitation of data in a more appropriate way [24]. In fact, the use of the intermediate metamodel is a benefit itself, since it provides a standard way to access business-process data and also improves the interoperability among organizations.

The paper is organized as follows: Firstly, Sect. 2 gives a general overview of the approach and introduces how the proposed Business Instance Metamodel may be employed in different contexts. Secondly, Sect. 3 describes a case study that has been used to test the approach. Thirdly, the metamodel is detailed in Sect. 4. Section 5 presents the tool implemented as a proof of concept. The tool helps to define the matching between an Oracle^{TM} database and the Process Instance Metamodel. Section 6 then surveys other existing approaches that exploit data in different contexts. Finally, conclusions are drawn and further work is outlined in Sect. 7.

2   Approach Overview and Contributions

Business process data exploitation depends highly on the technology that supports the business data storage as well as how the data is structured. As a consequence, no standard approach exists that can exploit any type of source, and it is therefore necessary to develop ad-hoc data analysis mechanisms adapted to both data technology and model. Thus, for example, the necessary data preparation to generate an event log to be employed by a process-mining tool is totally different depending on whether data is stored in a relational database or whether it comes from a cloud data source or a data warehouse. Moreover, this generation process also depends on the specific data model.

In order to render data exploitation as technology-agnostic regarding its data structure [31], our approach is inspired by the guidelines provided by Model-Driven Architecture in such a way that we propose a Business Process Instance Metamodel that allows us to separate produced data structure from data analysis solutions. In other words, the Business-Process Instance Metamodel can be seen as an intermediate artifact that allows applications that produce business process data to become independent from those applications that consume such data [15]. Figure 1 depicts how the process instance metamodel acts as an interface between data producers and consumers.

The following subsections describe how the metamodel proposed in the paper could be used in different contexts under the previous viewpoints.

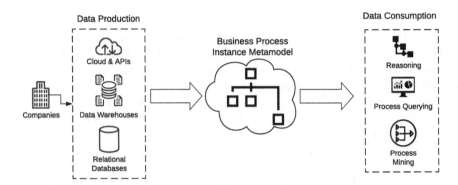

Fig. 1. Approach overview

2.1   Data Production Viewpoint

This viewpoint represents those contexts in which business-process data is produced. This data is mapped into the metamodel in order to be analysed. The left-hand side of Fig. 1 depicts three different contexts of use related to formatting the produced data: APIs located in cloud systems; data warehouses; and relational databases.

Regarding the context of cloud systems and APIs, it should be borne in mind that companies use more and more cloud data sources which rely on complex structures such as JSON, whose objects might have different properties. This data usually complements the specific company data with data from payments, geolocation, etc. Furthermore, APIs can be used to cross-reference information (such as weather and macroeconomics) and cloud systems usually perform some computation over data, which results in new data sources.

Data warehouse based applications are an especially important context of use, above all when the business process data produced is used as input for process-mining and process-discovery techniques, since data warehouses commonly store historical information of the companies as well as many details regarding the timing of that information. Note that process-mining techniques need historical information in order to rebuild a consistent process model.

Finally, a common context of use from the data production viewpoint is related to applications that use relational databases. In fact, the case study introduced in this paper is based on a relational database. Relational databases also provide one of the most widely used scenarios for process querying, as detailed in Sect. 6.

2.2   Data Consumption Viewpoint

This viewpoint represents those contexts in which data obtained from business processes is exploited. The right-hand side of Fig. 1 depicts three different contexts of use related to data exploitation: reasoning by using defined ontologies;

process queries for the creation of dashboards that improve decision-making; and event log generation for process mining.

In the reasoning context of use, applications use semantization techniques, which are based on an ontology as a formal specification. Many business process semantization approaches link concepts from domain ontologies with business process elements that are grounded in a business process ontology [19]. Thus, reasoning is used to derive facts from the ontology, which are not expressed explicitly. The elements of this business process ontology can be mapped to the concepts defined in our metamodel, since ontologies and metamodels are closely related [29].

Process queries improve decision-making, for example, by creating dashboards to exploit the business process data [35]. As far as our metamodel covers information related to process definitions, process instances, activities and activity instances and their attributes, we can ask for durations, sequences of activities, frequencies of executions, and can identify bottleneck activities, study deviated instances of activities/processes, etc. As a consequence, this information can be used to infer Key Performance Indicators which facilitate the monitoring of the process [32].

Finally, in the context of event logs for process-mining techniques, applications may not be able to produce event logs or may fail to produce them in the correct format [2]. Obtaining logs from the instances of our metamodel implies listing the activity instances ordered in terms of execution time, grouping them by process instance, and producing files with XES-formatted data. Note that the process to perform this transformation must be adapted to the data source in order to obtain the correct data output for processing. Moreover, it must be considered that the company systems work with various data sources at the same time. From the consumer viewpoint, all these details must be transparent by means of an appropriate transformation.

3   Case Study

This section presents the case study carried out to test the validity of the proposal. Data has been obtained from the execution of a business process within a prominent aerospace company. Although the company has no Business Process Management System, it does have a proprietary system that is supported by a relational database. The core business of the company consists of the assembly of aircraft and their modules. An aircraft undergoes a huge process of engineering, components designing, components construction that must be followed to assemble the final product to be ready to fly. When a new aircraft is about to be released, it must be tested several times. Figure 2 depicts the testing process of the aircraft modules.

When the aircraft testing process starts, the *New aircraft order arrives* activity begins, and, as a consequence, its data is introduced in the AIRCRAFT table (Fig. 3). Bear in mind that an aircraft passes through different stations, and that in each station, the aircraft modules must pass a set of tests that are composed of

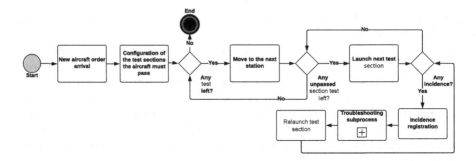

Fig. 2. Aircraft testing process

different sections. Thereby the execution of each test brings about the execution of every section in that test.

The *Configuration of the test sections that the aircraft must pass* activity consists of scheduling the different test sections that must be executed on the modules of each aircraft. This information is stored in the TEST_SECTIONS table. After the test configuration, the aircraft is driven to the first station to start the test execution (*Move to the next station* activity) and the set of tests are launched (*Launch next test* activity).

Every time a test is launched, a row is inserted in the TEST_EXECUTION table and another row is inserted in TEST_SECTION_EXECUTIONS (one for each test section executed). Furthermore, if the test fails, usually due to some kind of incidence, the *Incidence Registration* activity starts and the *Troubleshooting subprocess* is triggered. As a consequence: first, the original row is modified in order to register both the moment when the test failed and the status of the test after being executed; and second, a new row is inserted in the TEST_INCIDENCES table. Note that if an incidence appears during the execution of a section, it must be solved successfully before the airplane is released. As a consequence, the full test needs to be repeated, regardless of which section the error appeared, since the success of some parts of a test may depend on other parts of the test. Thus, when the *Troubleshooting subprocess* finishes, then the whole test is relaunched (*Relaunch test* activity).

Due to the lack of a Business Process Management System, every test execution is stored in detail in the database, whereby information related to aircraft, tests, stations, etc. is held. Thus, each time a new test is launched, the data involved is stored, such as timestamps related to every action, the status of the test, when the test has finished, and which sections were executed. The data model which supports this process is composed of the following tables (Fig. 3):

– AIRCRAFT: This stores information about the tested aircraft. As a consequence, a row is inserted in this table each time an airplane is going to be tested. The table stores: the type of the airplane, the model of the airplane, the name, the start date, and the end date scheduled.

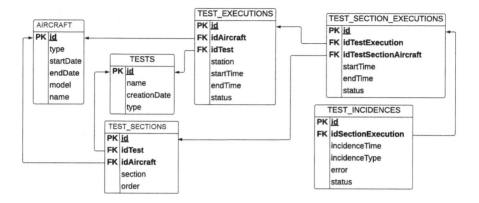

Fig. 3. Relational model of the aircraft assembly process

- TESTS: This stores information about the collection of tests defined in the system: the name of the test, the creation date, and its type.
- TEST_SECTIONS: This stores the sections of each test that each airplane should pass and the order in which the sections should be executed.
- TEST_EXECUTIONS: This table stores information about a test execution. Thus, a row is inserted in this table each time a test is launched. The table stores: the station in which the test was executed, the time when the test execution started, the time when the test execution ended, and the test status after finishing.
- TEST_SECTION_EXECUTIONS: This stores information about the execution of a test section. Note that each test is split into different sections that are in charge of preparing the execution or checking certain variables. The table stores the timestamp when the test section execution started, the timestamp when the test section execution ended, and the final status.
- TEST_INCIDENCES: This stores information about the incidences produced during test executions. As a consequence, a row is inserted in this table when an error appears while running a test. The table stores the time when the incidence appeared, the incidence type, the status of the incidence, and the error that caused the incidence.

4   Process Instance Metamodel

The Business Process Instance Metamodel is detailed in Fig. 4. The metamodel has been specified with EMF [37]. Note that it is a very simple model which is mainly centred on the most basic entities related to business process instances together with their attributes. A previous extension of this metamodel was published in [15].

The root of the metamodel is the **ProcessEngine** metaclass and represents the BPMS or software application that is in charge of process execution. The process engine can be in charge of several processes. The **ProcessDefinition** metaclass represents the formal definition of the process, that is, what we call the Business Process Model. The attributes are:

- *id.* Key identifier of the process.
- *name.* Name of the process model.
- *description.* Description of the process model.
- *suspended.* This attribute represents whether a process is suspended (temporarily disabled). While it is suspended, the process is not instantiated.

A business process is composed of different activities and the **Activity** metaclass models these activities. The attributes are:

- *id.* Key identifier of the activity.
- *name.* Name of the activity.
- *description.* Description of the activity.

One business process can be executed many times and the **ProcessInstance** metaclass models these executions or instances. The attributes are:

- *id.* Key identifier of the process instance.
- *ended.* A flag (Boolean) indicating whether the instance is still running.
- *suspended.* A flag (Boolean) indicating whether the instance is suspended.
- *startUser.* The user who started the instance process.
- *duration.* Time spent on process execution. This information is recovered when the process has ended.
- *startTime.* This represents when the instance process started.
- *endTime.* This represents when the instance process ended.

Finally, the **ActivityInstance** metaclass represents the execution of an activity and is related to the **Activity** metaclass (note that an activity may be executed many times) and to the **ProcessInstance** metaclass (an activity may be executed in the context of different business processes). The attributes are:

- *id.* Key identifier of the activity instance.
- *startTime.* This represents when the instance activity started.
- *endTime.* This represents when the instance activity ended.
- *duration.* Time spent on activity execution. This information is recovered when the activity ends.
- *cancelled.* A flag (Boolean) indicating whether the instance is cancelled.
- *assignee.* The user assigned to the execution of the activity.

Note that this metamodel allows us to exploit business data in different contexts, independently of the storage technology and how the information is structured. We only need to define mappings from the concrete technology to the Process Instance Metamodel. Therefore, the information stored as instances of the Business Process Instance Metamodel may be used to generate event log traces (both in XES or MXML format), to be queried for decision-making or to be semantized thereby enabling the application of reasoning techniques.

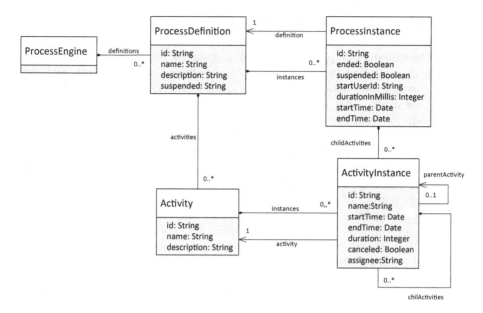

Fig. 4. Process Instance Metamodel

4.1 Mapping the Metamodel and the Case Study Models

This section explains how the Process Instance Metamodel is used, from the data production viewpoint, in the context of the case study introduced in Sect. 3.

The **Process Definition** metaclass is related to the *Testing Aircraft Process* shown in Fig. 2. Each instance of that process is mapped into the **Process Instance** metaclass (see Fig. 5). As a consequence, when a new row is inserted into the TEST_EXECUTIONS table, then an instance of the **Process Instance** metaclass is created. It should be also noted that the *startTime* and *endTime* attributes are mapped to the *startTime* and *endTime* fields of that table.

Since there are different activities, such as the *Launch next test* or the *Incidence registration* activities, there are mappings between the **Activity Instance** metaclass and different tables (see Fig. 5). Thus, an instance of the **Activity Instance** metaclass is created each time a new row is inserted into the TEST_SECTION_EXECUTIONS, TEST_INCIDENCES, or TEST_EXECUTIONS tables. The expected *startDate* of an assembly process of an airplane is stored in the AIRCRAFT table. However, the real start time is represented by the oldest *startTime* of the TEST_EXECUTION related to a specific *idAircraft* that indicates the true beginning of the process.

Finally note that, although in this case study every mapping is related to the insertion of a row into a table, this is not the only possible scenario. The mappings of the *Activity Instance* metaclass could also be related to editions of rows. Thus, for example, the *incidenceType* field could be mapped to different activities if the various types of incidences lead to different subprocess executions.

5   Proof of Concept Implementation

In order to support our proposal, a proof-of-concept has been implemented to illustrate the mapping process between the business data repository and our Process Instance Metamodel. One of the main benefits of our proposal is that the mapping itself is that which remains after performing the matching between data repositories and the metamodel, instead of the mapped data, as in other approaches [7]. Thus, every new item of data registered in the repository is auto-matically available in the Process Instance Metamodel, and it is able to perform business process analysis not just after the process execution, but also whilst the execution is happening, which is key in some cases. This provides agility and the opportunity of making decisions during the business process instance. Another considerable benefit of this approach, since it is not tied to any spe-cific data consumption context (process mining, process querying, reasoning over processes...), is the ability to exploit the business process data simultaneously in different contexts. We could use it to generate event logs while visualizing statistic data on a dashboard, as is showed in the demo video recorded using the proof-of-concept tool. This provides versatility to the way business process data can be used by the companies without the necessity of performing specific ad-hoc applications or data transformations for each context or goal. The proof of concept has been developed as a web application and implements a simple dash-board where we can compare visually different instances, cross-check statistics information related to our instances, and watch the evolution of our process data over time. Furthermore, a video demo shows how the tool is able to automati-cally analyse the structure of the data repository and how the mapping process can be executed in an easy way. Figure 6 shows a screenshot that captures the mapping definition process. The proof-of-concept has been developed as a result of collaboration with a company whose data could never be publicly available. However, the software is available for application to other cases. Moreover, a video demo using the tool has been recorded to facilitate the use of the tool. For any further details regarding the tool, check the website http://www.idea.us.es/portfolio-item/process-data-matching-tool/.

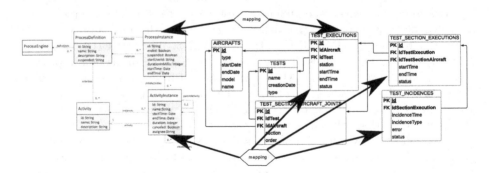

Fig. 5. Mappings between metamodel and produced data

6   Related Work

We will limit the scope of this section to the approaches related to business processes whose focus is on the exploitation of data generated during process execution. The approaches can be classified into: approaches whose goal is the semantization of process data in order to use ontology-based reasoning; approaches whose goal involves the querying of process data to aid in decision-making in business process scenarios; and approaches whose goal is the creation of execution traces that are used as input for process discovery algorithms. Bear in mind that these different scenarios consume data in different formats, and certain conversion and formatting tasks can be tedious and complex since data can be stored in heterogeneous repositories [7]. The following subsections give a general overview of the state-of-the-art of the aforementioned contexts.

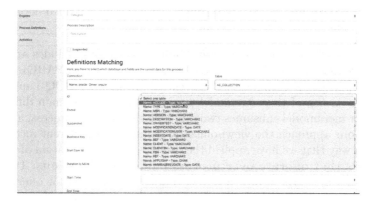

Fig. 6. Developed tool as proof of concept

6.1   Approaches that Consume Data for Reasoning

This group introduces the incorporation of data ontology in order to support functionalities of a more intelligent nature, such as process reasoning. In general, these approaches augment existing processes with semantic annotations, so that formal reasoning techniques can be applied. There are several techniques of semantization of Business Processes [20]:

- The SUPER project [40] formally represents business process concepts by means of a stack of five ontologies and provides a modelling environment for the enrichment of existing processes with semantic annotations.
- The SAP AG system [6] integrates semantic descriptions and business process artifacts by linking concepts from an ontology and elements of business process models.
- The Prosecco project [28] provides a unified dictionary of business concepts to help with the systems integration and takes into account semantic dependencies between business process models and rule models.

- Finally, there is also a group of techniques that could be used for semantization of business processes that are not process specific, for example, since many business process execution environments use REST interfaces, certain techniques for semantization of REST interfaces could be used. However, these kinds of techniques remain out of the scope of this study.

The approach that is most closely related to ours is the SAP AG system in the sense that the domain ontology and the business process model are integrated by means of links; however, that system is focused on semantic sources.

6.2   Approaches that Consume Data for Querying

The approaches in this group query process data to help in decision-making in business process scenarios [35]. There are many different approaches to query process data. According to [33], these approaches can be classified depending on the type of behaviour models they can take as input:

- **Methods that operate over event logs.** This group includes approaches such as CRG [23], eCRG [21], DAPOQ-Lang [27], FPSPARQL [5], and PIQL [30]. The approach most closely related to ours is DAPOQ-Lang because it is built on top of the metamodel proposed in [26]. The main difference is that their metamodel subsumes two different viewpoints, (process, and data), while in our approach the viewpoints are defined with different metamodels, in such a way that we have applied the principle of the separation of concerns.
- **Methods that operate over process model specifications.** This group includes a set of approaches that were originally conceived for querying conceptual models, and, as a consequence, they are also useful for querying process models, and another set of approaches that were originally conceived for querying process model collections. The first subgroup includes approaches such as DMQL [11], GMQL [10], and VMQL [39]. The second subgroup includes approaches such as BPMN-Q [3], BPMN VQL [12], BPSL [22], CRL [13], Descriptive PQL [18], IPM-PQL [9], and PPSL [14]. The approaches most closely related to ours are DMQL and GMQL in the sense that they define a generic metamodel to cover all types of modelling languages. This metamodel can be seen as a way to decouple query languages from modelling languages.
- **Methods that operate over behaviours encoded in process models.** This group includes approaches such as APQL [16], BQL [17], QuBPAL [36], and PQL [34]. All of these approaches are based on the definition of semantic relations between tasks. The most closely related to our approach is that of APQL, in the sense that the proposed language is independent of the notation used to specify process models.
- **Methods that operate over collections that may include process models and/or event logs.** This group includes approaches such as BPQL [25] and NP-QL [4]. This group is the least related to our approach.

6.3   Approaches that Consume Data for the Creation of Execution Traces

There are several approaches that consume data to create execution traces. Thus, in [7], a conversion from a data source in table format to an event log is proposed. The approach is tested by means of two case studies: an SAP system, and a set of CSV files that are the result of exporting a database.

In [8], a framework to extract XES event log information from legacy relational databases is proposed. The extraction is made by defining two ontologies, one that represents the domain of interest and another one that represents event logs. The domain ontology is linked to the legacy data by using the ontology-based data access paradigm (OBDA), and the concepts defined in the event log ontology are mapped into the concepts defined in the domain ontology by means of annotations.

In [1], a framework to unify existing approaches of process discovery from event logs is introduced. The framework is based on event log and process model abstractions, and, as a consequence, it only includes concepts from event log and process viewpoints.

In [26], a metamodel is proposed to query the data from different sources in a standardized way. Thus, the metamodel allows the decoupling of the application of the data analysis techniques. The proposed metamodel includes concepts related to two different viewpoints: process and data. Furthermore, in order to be compatible with the XES metamodel, the proposed metamodel also includes events and cases. Mappings from data sources of three different scenarios (database redo logs, in-table version storage, and SAP-style change tables) to the proposed metamodel are formalized.

Note that all these approaches share at least one of the following two weak points covered by our approach: (1) different data sources are considered, but relational databases and/or tabular formats are taken for granted [7,26]; and (2) the focus is on the results (event logs) instead of on the means (relations between stored data and events data), which forces the process of mapping to be repeated each time a new log needs to be generated [1,7,8].

7   Conclusions and Further Work

Due to the existence of multiple techniques based on Business Process Analysis, this paper introduces the necessity of the utilization of a Business Process Instance Metamodel as a bridge between data sources and data exploitation techniques.

As we have seen, this metamodel provides the first step towards isolating the process data produced and the objective of its analysis. This is especially relevant in scenarios where different types of business process data exploitation are going to be applied and/or scenarios where different data sources with various formats are working together. Thereby, this paper shows how the use of an intermediate metamodel can help to standardize the exploitation of business process data by defining a common infrastructure that may be used in various contexts of business process analytics.

In terms of further work, how to query the metamodel in order to extract the required information in the correct format constitutes the next challenge to tackle. This challenge brings about the extension of the metamodel to encapsulate other existing proposals of consumers and producers, while it maintains the abstraction level to ensure adaptability to any business regardless of its sector or domain knowledge.

Finally, we consider it interesting to enrich the way of defining the matching, by making the tool more flexible and by allowing the building of processes of a more complex nature and the exploitation of more complex data sources.

Acknowledgements. This work has been partially funded by the Ministry of Science and Technology of Spain by the Project ECLIPSE (RTI2018-094283-B-C33 and TIN2016-75394-R) and the European Regional Development Fund (ERDF/FEDER).

References

1. van der Aalst, W.M.P.: Process discovery from event data: relating models and logs through abstractions. Wiley Interdisc. Rew.: Data Min. Knowl. Discov. **8**(3) (2018). https://doi.org/10.1002/widm.1244
2. van der Aalst, W.M.P.: Process mining: the missing link. Process Mining, pp. 25–52. Springer, Heidelberg (2016). https://doi.org/10.1007/978-3-662-49851-4_2
3. Awad, A.: BPMN-Q: a language to query business processes. In: Enterprise Modelling and Information Systems Architectures, pp. 115–128 (2007)
4. Beeri, C., Eyal, A., Kamenkovich, S., Milo, T.: Querying business processes. In: Proceedings of the 32nd International Conference on Very Large Data Bases. VLDB 2006, pp. 343–354. VLDB Endowment (2006). http://dl.acm.org/citation. cfm?id=1182635.1164158
5. Beheshti, S.-M.-R., Benatallah, B., Motahari-Nezhad, H.R., Sakr, S.: A query language for analyzing business processes execution. In: Rinderle-Ma, S., Toumani, F., Wolf, K. (eds.) BPM 2011. LNCS, vol. 6896, pp. 281–297. Springer, Heidelberg (2011). https://doi.org/10.1007/978-3-642-23059-2_22
6. Born, M., Dörr, F., Weber, I.: User-friendly semantic annotation in business process modeling. In: Weske, M., Hacid, M.-S., Godart, C. (eds.) WISE 2007. LNCS, vol. 4832, pp. 260–271. Springer, Heidelberg (2007). https://doi.org/10.1007/978-3-540-77010-7_25
7. Buijs, J.: Mapping data sources to XES in a generic way. Department of Mathematics and Computer Science, Eindhoven University of Technology (2010)
8. Calvanese, D., Montali, M., Syamsiyah, A., van der Aalst, W.M.P.: Ontology-driven extraction of event logs from relational databases. In: Reichert, M., Reijers, H.A. (eds.) BPM 2015. LNBIP, vol. 256, pp. 140–153. Springer, Cham (2016). https://doi.org/10.1007/978-3-319-42887-1_12
9. Choi, I., Kim, K., Jang, M.: An XML-based process repository and process query language for integrated process management. Knowl. Process Manag. **14**(4), 303–316. https://onlinelibrary.wiley.com/doi/abs/10.1002/kpm.290
10. Delfmann, P., Breuker, D., Matzner, M., Becker, J.: Supporting information systems analysis through conceptual model query. The diagramed model query language (DMQL). Commun. Assoc. Inf. Syst. **37**, 24 (2015)

11. Delfmann, P., Steinhorst, M., Dietrich, H.A., Becker, J.: The generic model query language GMQL. Conceptual specification, implementation, and runtime evaluation. Inf. Syst. **47**, 129–177 (2015)
12. Di Francescomarino, C., Tonella, P.: Crosscutting concern documentation by visual query of business processes. In: Ardagna, D., Mecella, M., Yang, J. (eds.) BPM 2008. LNBIP, vol. 17, pp. 18–31. Springer, Heidelberg (2009). https://doi.org/10.1007/978-3-642-00328-8_3
13. Elgammal, A., Turetken, O., van den Heuvel, W.-J., Papazoglou, M.: Formalizing and appling compliance patterns for business process compliance. Softw. Syst. Model. **15**(1), 119–146 (2014). https://doi.org/10.1007/s10270-014-0395-3
14. Foerster, A., Engels, G., Schattkowsky, T.: Activity diagram patterns for modeling quality constraints in business processes. In: Briand, L., Williams, C. (eds.) MODELS 2005. LNCS, vol. 3713, pp. 2–16. Springer, Heidelberg (2005). https://doi.org/10.1007/11557432_2
15. Gómez-López, M.T., Reina Quintero, A.M., Parody, L., Pérez Álvarez, J.M., Reichert, M.: An architecture for querying business process, business process instances, and business data models. In: Teniente, E., Weidlich, M. (eds.) BPM 2017. LNBIP, vol. 308, pp. 757–769. Springer, Cham (2018). https://doi.org/10.1007/978-3-319-74030-0_60
16. ter Hofstede, A.H.M., Ouyang, C., La Rosa, M., Song, L., Wang, J., Polyvyanyy, A.: APQL: a process-model query language. In: Song, M., Wynn, M.T., Liu, J. (eds.) AP-BPM 2013. LNBIP, vol. 159, pp. 23–38. Springer, Cham (2013). https://doi.org/10.1007/978-3-319-02922-1_2
17. Jin, T., Wang, J., Wen, L.: Querying business process models based on semantics. In: Yu, J.X., Kim, M.H., Unland, R. (eds.) DASFAA 2011. LNCS, vol. 6588, pp. 164–178. Springer, Heidelberg (2011). https://doi.org/10.1007/978-3-642-20152-3_13
18. Kammerer, K., Kolb, J., Reichert, M.: PQL - a descriptive language for querying, abstracting and changing process models. In: Gaaloul, K., Schmidt, R., Nurcan, S., Guerreiro, S., Ma, Q. (eds.) CAISE 2015. LNBIP, vol. 214, pp. 135–150. Springer, Cham (2015). https://doi.org/10.1007/978-3-319-19237-6_9
19. Kluza, K., Kaczor, K., Nalepa, G.J., Ślażyński, M.: Opportunities for business process semantization in open-source process execution environments. In: 2015 Federated Conference on Computer Science and Information Systems (FedCSIS), vol. 5, pp. 1307–1314, September 2015
20. Kluza, K., et al.: Overview of selected business process semantization techniques. In: Pełech-Pilichowski, T., Mach-Król, M., Olszak, C.M. (eds.) Advances in Business ICT: New Ideas from Ongoing Research. SCI, vol. 658, pp. 45–64. Springer, Cham (2017). https://doi.org/10.1007/978-3-319-47208-9_4
21. Knuplesch, D., Reichert, M.: A visual language for modeling multiple perspectives of business process compliance rules. Softw. Syst. Model. **16**(3), 715–736 (2016). https://doi.org/10.1007/s10270-016-0526-0
22. Liu, Y., Muller, S., Xu, K.: A static compliance-checking framework for business process models. IBM Syst. J. **46**(2), 335–361 (2007)
23. Ly, L.T., Rinderle-Ma, S., Dadam, P.: Design and verification of instantiable compliance rule graphs in process-aware information systems. In: Pernici, B. (ed.) CAiSE 2010. LNCS, vol. 6051, pp. 9–23. Springer, Heidelberg (2010). https://doi.org/10.1007/978-3-642-13094-6_3
24. Mannhardt, F., de Leoni, M., Reijers, H.A., van der Aalst, W.M.P., Toussaint, P.J.: Guided process discovery - a pattern-based approach. Inf. Syst. **76**, 1–18 (2018). https://doi.org/10.1016/j.is.2018.01.009

25. Momotko, M., Subieta, K.: Process query language: a way to make workflow processes more flexible. In: Benczúr, A., Demetrovics, J., Gottlob, G. (eds.) ADBIS 2004. LNCS, vol. 3255, pp. 306–321. Springer, Heidelberg (2004). https://doi.org/10.1007/978-3-540-30204-9_21

26. González López de Murillas, E., Reijers, H.A., van der Aalst, W.M.P.: Connecting databases with process mining: a meta model and toolset. Softw. Syst. Model. **18**(2), 1209–1247 (2018). https://doi.org/10.1007/s10270-018-0664-7

27. González López de Murillas, E., Reijers, H.A., van der Aalst, W.M.P.: Everything you always wanted to know about your process, but did not know how to ask. In: Dumas, M., Fantinato, M. (eds.) BPM 2016. LNBIP, vol. 281, pp. 296–309. Springer, Cham (2017). https://doi.org/10.1007/978-3-319-58457-7_22

28. Nalepa, G.J., Ślażyński, M., Kutt, K., Kucharska, E., Łuszpaj, A.: Unifying business concepts for SMEs with prosecco ontology. In: 2015 Federated Conference on Computer Science and Information Systems (FedCSIS), pp. 1321–1326, September 2015

29. Parreiras, F.S.: Marrying ontology and model-driven engineering (chap.). Wiley and Sons (2012)

30. Pérez-Álvarez, J.M., Gómez-López, M.T., Parody, L., Gasca, R.M.: Process instance query language to include process performance indicators in DMN. In: 2016 IEEE 20th International Enterprise Distributed Object Computing Workshop (EDOCW), pp. 1–8, September 2016

31. Pérez-Álvarez, J.M., Gómez-López, M.T., Eshuis, R., Montali, M., Gasca, R.M.: Verifying the manipulation of data objects according to business process and data models. Knowl. Inf. Syst., 1–31 (2020). https://doi.org/10.1007/s10115-019-01431-5

32. Pérez-Álvarez, J.M., Maté, A., Gómez-López, M.T., Trujillo, J.: Tactical business-process-decision support based on KPIs monitoring and validation. Comput. Ind. **102**, 23–39 (2018). https://doi.org/10.1016/j.compind.2018.08.001

33. Polyvyanyy, A.: Business process querying. In: Sakr, S., Zomaya, A. (eds.) Encyclopedia of Big Data Technologies, pp. 1–9. Springer, Cham (2018). https://doi.org/10.1007/978-3-319-63962-8_108-1

34. Polyvyanyy, A., Corno, L., Conforti, R., Raboczi, S., Rosa, M.L., Fortino, G.: Process querying in Apromore. In: BPM 2015 Demo Track (2015)

35. Polyvyanyy, A., Ouyang, C., Barros, A., van der Aalst, W.M.P.: Process querying: enabling business intelligence through query-based process analytics. Decis. Support Syst. **100**, 41–56 (2017). https://doi.org/10.1016/j.dss.2017.04.011

36. Smith, F., Missikoff, M., Proietti, M.: Ontology-based querying of composite services. In: Ardagna, C.A., Damiani, E., Maciaszek, L.A., Missikoff, M., Parkin, M. (eds.) Business System Management and Engineering. LNCS, vol. 7350, pp. 159–180. Springer, Heidelberg (2012). https://doi.org/10.1007/978-3-642-32439-0_10

37. Steinberg, D., Budinsky, F., Paternostro, M., Merks, E.: EMF: Eclipse Modeling Framework 2.0, 2nd edn. Addison-Wesley Professional, Boston (2009)

38. Stodder, D.: Improving data preparation for business analytics. Applying technologies and methods for establishing trusted data assets for more productive users. Best Practices Report Q3 2016, pp. 19–21 (2016)

39. Störrle, H.: VMQL: a visual language for ad-hoc model querying. J. Vis. Lang. Comput. **22**(1), 3–29 (2011). Special Issue on Visual Languages and Logic

40. Wetzstein, B., et al.: Semantic business process management: a lifecycle based requirements analysis. In: Proceedings of the Workshop on Semantic Business Process and Product Lifecycle Management, vol. 251. CEUR Workshop Proceedings (2007)

Exploiting Event Log Event Attributes in RNN Based Prediction

Markku Hinkka[1,3](✉) (iD), Teemu Lehto[1,3] (iD), and Keijo Heljanko[2,4] (iD)

[1] School of Science, Department of Computer Science,
Aalto University, Espoo, Finland
`markku.hinkka@aalto.fi`
[2] Department of Computer Science,
University of Helsinki, Helsinki, Finland
`keijo.heljanko@helsinki.fi`
[3] QPR Software Plc, Helsinki, Finland
`teemu.lehto@qpr.com`
[4] HIIT Helsinki Institute for Information Technology,
Helsinki, Finland

Abstract. In predictive process analytics, current and historical process data in event logs is used to predict the future, e.g., to predict the next activity or how long a process will still require to complete. Recurrent neural networks (RNN) and its subclasses have been demonstrated to be well suited for creating prediction models. Thus far, event attributes have not been fully utilized in these models. The biggest challenge in exploiting them in prediction models is the potentially large amount of event attributes and attribute values. We present a novel clustering technique that allows for trade-offs between prediction accuracy and the time needed for model training and prediction. As an additional finding, we also find that this clustering method combined with having raw event attribute values in some cases provides even better prediction accuracy at the cost of additional time required for training and prediction.

Keywords: Process mining · Predictive process analytics · Prediction · Recurrent neural networks · Gated Recurrent Unit

1 Introduction

Event logs generated by systems in business processes are used in Process Mining to automatically build real-life process definitions and as-is models behind those event logs. There is a growing number of applications for predicting the properties of newly added event log cases, or process instances, based on case data imported earlier into the system [4,5,13,19]. The more the users start to understand their processes, the more they want to optimize them. This optimization can be facilitated by performing predictions. To be able to predict properties of new and ongoing cases, as much information as possible should be collected that

© IFIP International Federation for Information Processing 2020
Published by Springer Nature Switzerland AG 2020
P. Ceravolo et al. (Eds.): SIMPDA 2018/2019, LNBIP 379, pp. 67–85, 2020.
https://doi.org/10.1007/978-3-030-46633-6_4

is related to the event log traces and relevant to the properties to be predicted. Based on this information, a model of the system creating the event logs can be created. In our approach, the model creation is performed using supervised machine learning techniques.

In our previous work [9] we have explored the possibility to use machine learning techniques for classification and root cause analysis for a process mining-related classification task. In the paper, experiments were performed on the efficiency of several feature selection techniques and sets of structural features (a.k.a. activity patterns) based on process paths in process mining models in the context of a classification task. One of the biggest problems with the approach is the finding of the structural features having the most impact on the classification result. E.g., whether to use only activity occurrences, transitions between two activities, activity orders, or other even more complicated types of structural features such as detecting subprocesses or repeats. For this purpose, we have proposed another approach in [10], where we have examined the use of recurrent neural network techniques for classification and prediction. These techniques are capable of automatically learning more complicated causal relationships between activity occurrences in *activity sequences*. We have evaluated several different approaches and parameters for the recurrent neural network techniques and have compared the results with the results we collected in our work. In both the previous publications [9,10], focusing on boolean-type classification tasks based on the *activity sequences* only.

In this work we build on our previous work to further improve the prediction accuracy of prediction models by exploiting additional event attributes that are often available in the event logs while also taking into account the scalability of the approach to allow users to precisely specify the event attribute detail level suitable for the prediction task ahead. Our goal is to develop a technique that would allow the creation of a tool that is, based on a relatively simple set of parameters and training data, able to efficiently produce a prediction model for any case-level prediction task, such as predicting the next activity or the final duration of a running case. Fast model rebuilding is also required in order for a tool to be able to also support, e.g., interactive event and case filtering capabilities. Thus, the performance of the system under study is measured using four different metrics: Success rate, input vector length that gives a rough indication of the memory usage, the time required for training a model and the time required for performing predictions using the already trained model.

To answer these requirements, we introduce a novel method of exploiting event attributes into RNN prediction models by clustering events by their event attribute values and using the cluster labels in the RNN input vectors instead of the raw event data. This makes it easy to manage the input RNN vector size no matter how many event attributes there are in the data set. E.g., users can configure the absolute maximum length of the one-hot vector used for the event attribute data which will not be exceeded, no matter how many actual attributes the dataset has. RNN is an ideal choice for process mining-related prediction tasks since it learns temporal dynamic behavior by using its internal state to process sequences of inputs. Since predictions are usually made based on sequences of events, this makes it a more natural machine learning technique in process mining context than more traditional approaches, such as random forest and gradient boosting.

Our prediction engine source code is available in GitHub[1].

The rest of this paper is structured as follows: Sect. 2 is a summary of the latest developments around the subject. In Sect. 3, we present the problem statement and the related concepts. Section 4 presents our solution for the problem. In Sect. 5 we present our test framework used to test our solution. Section 6 describes the used datasets as well as performed prediction scenarios. Section 7 presents the experiments and their results validating our solution. Finally Sect. 8 draws the final conclusions.

2 Related Work

Lately, there has been a lot of interest in the academic world on predictive process monitoring which can clearly be seen, e.g., in [6] where the authors have collected a survey of 55 accepted academic papers on the subject. In [18], the authors have compared several approaches spanning three different research fields: Machine learning, process mining and grammar inference. As a result, they have found that overall, the techniques from machine learning field generate more accurate predictions than grammar inference and process mining fields.

In [19] the authors used Long Short-Term Memory (LSTM) recurrent neural networks to predict the next activity and its timestamp. They use one-hot encoded activity labels and three numerical time-based features: duration between the current activity and the previous activity, time within the day and time within the week. Event attributes were not considered at all. In [4] the authors trained LSTM networks to predict the next activity. In this case, however, network inputs are created by concatenating categorical, character string-valued event attributes and then encoding these attributes via an embedding space. They also note that this approach is feasible only because of the small number of unique values each attribute had in their test datasets. Similarly, in [17], the authors take a very similar approach based on LSTM networks, but this time also incorporate both discrete and continuous event attribute values. Discrete values are one-hot encoded, whereas continuous values are normalized using min-max normalization and added to the input vectors as single values.

In [14] the authors use Gated Recurrent Unit (GRU) recurrent neural networks to detect *anomalies* in event logs. One one-hot encoded vector is created for activity labels and one for each of the included string-valued event attributes. These vectors are then concatenated in a similar fashion to our solution into one vector representing one event, which is then given as input to the network. We use this approach for benchmarking our own clustering-based approach (labeled as *Raw* feature in the text below). The system proposed in their paper is able to predict both the next activity and the next values of event attributes. Specifically, it does not take case attributes and temporal attributes into account.

---

[1] https://github.com/mhinkka/articles.

In [21] the authors train a RNN to predict the most likely future activity sequence of a running process based only on the sequence of activity labels. Similarly our earlier publication [9] used sequences of activity labels to train a LSTM network to perform a boolean classification of cases.

Also, process mining models obtained using process mining techniques themselves can be used as a model for prediction. In [1] the authors first generate a process tree using the inductive miner algorithm, after which this process tree is used to predict the future path of ongoing cases. This approach does not take any additional event- or case attributes into account.

None of the mentioned earlier works present a solution that is scalable for datasets having lots of event- or case attributes and unique attribute values.

3 Problem

Using RNN to perform case-level predictions on event logs has lately been studied a lot. However, there has not been any scalable approach to handling event attributes in the RNN setting. Instead, e.g., in [14] authors used separate one-hot encoded vector for each attribute value. Having this kind of an approach when you have, e.g., 10 different attributes, each having 10 unique values would already require a vector of 100 elements to be added as input for every event. The longer the input vectors become, the more time and memory it gets for the model to create accurate models from them. This increases the time and memory required to use the model for predictions.

4 Solution

Since in addition to having activity labels in the input vectors, we need to store also event attribute-related information, we decided to include several feature types into the input vectors of the RNN. Input vectors are formatted as shown in Table 1, where each column represents one feature vector element f_{ab}, where a is the index of the feature and b is the index of the element of that feature. In the table, n represents the number of feature types used in the feature vector and m_k represents the number of elements required in the input vector for feature type k. Thus, each feature type produces one or more numeric elements into the input vector, which are then concatenated together into one actual input vector passed to RNN both in training and in prediction phases. Table 2 shows an example input vector having three different feature types: activity label, raw event attribute values (only single event attribute named *food* having four unique values) and the event attribute cluster where clustering has been performed separately for each unique activity.

For this paper, we encoded only event activity labels and event attributes into the input vectors. However, this mechanism can easily incorporate also other types of features not described here. The only requirement for added features is that it needs to be able to be encoded into a numeric vector as shown in Table 1 whose length must be the same for each event.

Table 1. Feature input vector structure

f_{11}	f_{12}	...	f_{1m_1}	f_{21}	...	f_{2m_2}	...	f_{n1}	...	f_{nm_n}

Table 2. Feature input vector example content

Row	$activity_{eat}$	$activity_{drink}$	$food_{salad}$	$food_{pizza}$	$food_{water}$	$food_{soda}$	$cluster_1$	$cluster_2$
1	1	0	1	0	0	0	1	0
2	0	1	0	0	1	0	1	0
3	1	0	0	1	0	0	0	1
4	1	0	0	1	0	0	0	1
5	0	1	0	0	0	1	0	1

4.1 Event Attributes

Our primary solution for incorporating information in event attributes into input vectors is to cluster all the event attribute values in the training set and then use a one-hot encoded cluster identifier to represent all the attribute values of the element. The used clustering algorithm must be such that it tries to automatically find the optimal number of clusters for the given data set within the range of 0 to N clusters, where N can be configured by the user. By changing N, the user can easily configure the maximum length of the one-hot -vector as well as the precision of how detailed attribute information will be tracked. For this paper, we experimented with a slightly modified version of Xmeans-algorithm [15]. Another option could have been a method where silhouette scoring [16] is used to determine the optimal number of clusters for k-means [8], but based on our tests, this approach did not perform fast enough to be applied as our selected approach.

It is very common that different activities get processed by different resources yielding a completely different set of possible attribute values. E.g., different departments in a hospital have different people, materials and processes. Also, in the example feature vector shown in Table 2, *food* -event attribute has completely different set of possible values depending on the *activity* since it is forbidden by, e.g., the external system to not allow activity of type *eat* to have *food* event attribute value of *water*. If we cluster all the event attributes using single clustering, we would easily lose this activity type-specific information.

In order to retain this activity-specific information, we used separate clustering for each unique activity type. All the event attribute clusters are encoded into one one-hot encoded vector representing only the resulting cluster label for that event, no matter what its activity is. This is shown in the example table as $cluster_N$, which represents the row having N as a clustering label. E.g., in the example case, $cluster_1$ is 1 in both rows 1 and 2. However, row 1 is in that cluster because it is in the 1st cluster of the $activity_{eat}$ activity, whereas row 2 is in that cluster because it is in the 1st cluster of the $activity_{drink}$ activity. Thus, in order to identify the actual cluster, one would require both the activity label and the cluster label. For RNN to be able to properly learn about the actual event attribute values, it needs to be given both the activity label and the cluster label in the input vector. Below, this approach is labeled as *ClustN*, where N is the maximum cluster count.

For benchmarking, we also experimented with a *raw* implementation where event attributes were used so that every event attribute is encoded into its own one-hot encoded vector and then concatenated into the actual input vectors. This method is lossless since every unique event attribute value has its own indicator in the input vector. Below, this approach is referred to as *Raw*. Finally, we experimented also using both *Raw* and *Clustered* event attribute values. Below, this approach is referred to as *BothN*, where N is the maximum cluster count.

4.2   Formal Problem Definition

Basically, the problem we are solving in this paper is that we encode as much information as possible from event log event attributes into a single numeric RNN input vector whose length is user-configurable. In this section, we will present the formal definitions required to describe our clustering-based approach for solving this problem. We will build our formal definitions on the basis of the definitions given in Process Mining [20] -book Chapter 5 describing the basic concepts in process mining.

First, we define relationships between event log and events as follows:

Definition 1. *Event log L is a set of cases $c_i \in L$. Each case c_i is a sequence of events $c_i = \langle e_1, e_2, ..., e_n \rangle$. We denote $e \in L$ iff $\exists c_i = \langle e_1, e_2, ..., e_n \rangle \in L$, such that $e_j = e$, for some $1 \le j \le n$.*

Next, we define a function for accessing event attributes of an event:

Definition 2. *Let $\#_n : e \mapsto v$, where $(n \in AN) \land (e \in L)$, and AN is the finite set of all the attribute labels for the events in the given event log. v is the value of that attribute and can be of any arbitrary type.*

Using this function, we can refer to any of the *standard attributes* (in this paper, only *activity*, *time* and *caseid* are considered as standard attributes), as well as event log-specific attributes. For this paper, we need to define these attribute label sets as follows:

Definition 3. *Let $AN_{std} \subseteq AN$, be the set of standard attribute labels: $\langle activity, time, caseid \rangle$. Also, let $(AN_L \cap AN_{std} = \emptyset) \land (AN_{std} \cup AN_L = AN)$, where AN_L is the set of event attributes in event log L, which are not part of standard attributes.*

The standard attributes listed above have the following meanings:

Definition 4. *$\#_{activity}(e)$ is the activity label associated with the event e. This describes what has occurred. Activity labels in this paper are considered to be textual descriptions of the performed task. $\#_{time}(e)$ is the timestamp of the event e.*

This describes when something has occurred. In this paper, timestamps include both time and date of the occurred event.
$\#_{caseid}(e)$ *is the identifier of the case associated with the event e. This describes the (case) context of the occurrence. Case identifiers in this paper are considered to be short textual or numeric identifiers identifying the case.*

Using these, we can formally define a set of all the activity labels in the event log as follows:

Definition 5. *Let $\mathcal{A}_L$ be the set of activity labels so that $\forall e \in L, \#_{activity}(e) \in \mathcal{A}_L$.*

Next, we split the event log into two disjoint sets: Training set and test set. Formally this can be expressed as:

Definition 6. *Training set is $L_{tr} \subset L$, where $L_{tr} \neq \emptyset$. Similarly, test set is $L_t \subset L$, where $(L_t \neq \emptyset) \wedge (L_t \cap L_{tr} = \emptyset) \wedge (L_t \cup L_{tr} = L)$.*

We also formally define separate subsets of activity labels for both the test and the training set as follows:

Definition 7. *$\mathcal{A}_{tr} \subseteq \mathcal{A}_L$ is used to denote all the activity labels available in the training set, whereas $\mathcal{A}_t \subseteq \mathcal{A}_L$ denotes those in the test set.*

Similarly, we denote sets of available attributes in both the training set and the test set as follows:

Definition 8. *$AN_{tr} \subseteq AN_L$ is used to denote all the attribute names available in the training set, whereas $AN_t \subseteq AN_L$ denotes those in the test set.*

Next, we define a function used to concatenate multiple vectors to each other.

Definition 9. *Let $concat : \langle X_0, ..., X_n \rangle \mapsto Y$ be a function that returns a single numeric vector Y consisting of the concatenated contents of numeric vectors $X_0, ..., X_n$ in the specified order.*

Then we define a function that maps each unique value into unique integer value as follows:

Definition 10. *Let $codify : x \times X \mapsto y$, where X is a finite set of all the possible values for $x \in X$, and $y \in 1, ..., |X|$ is the value x being mapped bijectively into an integer. Thus, codify creates a bijection between values and an integer representation of that value.*

Now we can give a formal definition for the one-hot encoding function as follows:

Table 3. One-hot encoding example

Original	After *codify*	After *onehot*
$\langle a, b, c, d \rangle$	$\langle 1, 2, 3, 4 \rangle$	$\langle \langle 1,0,0,0 \rangle, \langle 0,1,0,0 \rangle, \langle 0,0,1,0 \rangle, \langle 0,0,0,1 \rangle \rangle$

Definition 11. *Let onehot* $: x \times X \mapsto Y$, *where X is a finite set of all the possible values and $x \in X$ is the value being encoded, be an onehot encoding function that transforms x into a numeric vector Y of length $|X|$. Vector $Y = \langle y_1, ..., y_{|X|} \rangle$, where*

$$y_k = \begin{cases} 1, & \text{iff } k = \text{codify}(x, X) \\ 0, & \text{otherwise} \end{cases}$$

Thus, every unique value in X returns an unique numeric vector.

An example of a one-hot encoding process is shown in Table 3, where the set of all possible values X is $\langle a, b, c, d \rangle$, and we are applying *onehot* function to each of the four possible values separately.

Next, since we need to be able to iterate through all the attributes, we need to specify an unambiguous way to map iteration index to an attribute name. For this purpose we define the following function:

Definition 12. *Let attname* $: n \mapsto an$, *where $an \in AN_L$, and $n \in 1, ..., |AN_L|$ be a bijective function for mapping an positive integer iteration index to an attribute name.*

Using the previous definition, we can refer to nth event attribute of e by writing $\#_{attname(n)}(e)$. Next, we define a method for creating subsets of events in a way that each subset will have all the events having one specific activity label. Formally we express these sets as:

Definition 13. *Let B_{act}, where $act \in \mathcal{A}_{tr}$ be the set of events $e \in L$ such that $\#_{activity}(e) = act$.*

Next, we will define a function for retrieving a set of all attribute values of a given set of events:

Definition 14. *Let $\#_n : \langle e_1, ..., e_n \rangle \mapsto Y$, where $e_i \in L \forall (1 \le i \le n)$. This function returns a set of all the attribute values of attribute having index n for given events.*

Using these definitions, the set of all the attribute values of all the events having a specific activity label act can be referred to using $\#_{attname(n)}(B_{act})$, which yields a set of attribute values of attribute having index n for all the events having activity label act. Next, In order to perform clustering for events in activity buckets, we need to first one-hot encode event attribute values.

Definition 15. *Let* $onehot_{B_{act}} : e \times n \mapsto Y$, *where* $e \in L$, $act \in \mathcal{A}_{tr}$, *and* $n \in 1, ..., |AN_L|$, *be a function that performs one-hot encoding for attribute having attribute iteration index of n for event e. Thus,*
$$Y = onehot(\#_{attname(n)}(e), \#_{attname(n)}(B_{\#_{activity}(e)})).$$

Using this definition, we can transform all event attribute values into a single numeric vector using the following additional function definition.

Definition 16. *One-hot encoding function for all event attributes of given event:*

$$onehot_\# : e \mapsto concat(onehot_{B_{\#_{activity}(e)}}(e, 1),$$

$$...,$$

$$onehot_{B_{\#_{activity}(e)}}(e, |AN_L|)),$$

where $e \in L$.

Next, we define the actual clustering function that uses the number vectors created by $onehot_\#$ function as follows.

Definition 17. $\#_{cluster} : e \times max_{cc} \mapsto y$, *where* $e \in L_{tr}$, *and* max_{cc} *denotes the configured maximum cluster count, and* $1 \leq y \leq max_{cc}$, *and* $y \in \mathbb{Z}$, *denotes the assigned cluster label among the set of all the possible cluster labels. Clustering is performed separately for each activity label* B_{act}, *where* $act \in \mathcal{A}_{tr}$ *using suitable clustering algorithm. Every clustered event is translated into clustering input vector using* $onehot_\#(e)$ *function. These input vectors are then used as data points for the clustering algorithm.*

Thus, every activity will have its own independent clustering having max_{cc} as the maximum cluster count. In the testing phase, the clusterings created from the training data will be used to fit the input vectors created from the events in the testing data.

Finally, we can specify the vector used as input vector in RNN training as follows:

Definition 18. *Generating input vector for one event in training is performed using:*

$$inputvector : e \times max_{cc} \mapsto concat(onehot(\#_{activity}(e), ACT_{tr}),$$

$$onehot(\#_{cluster}(e, max_{cc}), cl)),$$

where $e \in L_{tr}$, *and* $cl = \langle 1, ..., n \rangle$, *where* $n \leq max_{cc}$ *represents the set of all the possible cluster labels. The value of* $n \in \mathbb{Z}$ *depends on the clustering algorithm and represents the actual maximum number of clusters that were created for event attribute data of any single activity in* L_{tr}.

These input vectors are then passed to the RNN training as ordered sequences of event input vectors, where each sequence represents all the events of a single case in the L_{tr} in the order determined by their ascending timestamps.

As a result, when training, every event is preprocessed by performing the input vector generation using *inputvector* function. At the testing phase, the same encoding functions are used, however, if the event used in testing has some attributes, attribute values or activities that were not part of the training data set, those will just be ignored. Also, event attribute clustering in the testing phase is performed using the clustering models created in the training phase. Thus, in order to store the trained model, also all the trained clustering models must be stored.

5   Test Framework

We have performed our test runs using an extended Python-based prediction engine that was used in our earlier work [9]. The engine is still capable of supporting most of the hyperparameters that we experimented with in our earlier work, such as used RNN unit type, number of RNN layers and the used batch size. The prediction engine we built for this work takes a single JSON configuration file as input and outputs test result rows into a CSV file.

Tests were performed using a commonly used 3-fold cross-validation technique to measure the generalization characteristics of the trained models. In 3-fold cross-validation, the input data is split into three subsets of equal size. Each of the subsets is tested one by one against models trained using the other two subsets.

5.1   Training

Training begins by loading the event log data contained in the two of the three event log subsections. After this, the event log is split into actual training data and validation data that used to find the best performing model out of all the model states during all the test iterations. For this, we picked 75% of the cases for the training and the rest for the validation dataset. After this, we initialize event attribute clusters as described in Sect. 4.1.

The actual prediction model and the data used to generate the actual input vectors is performed next. This data initialization involves splitting cases into prefixes and also taking a random sample of the actual available data if the amount of data exceeds the configured maximum amount of prefixes. To avoid running out of memory during any of our tests, these limits were set to 75000 for training data and 25000 for validation data. We also had to filter out all the cases having more than 100 events.

Finally, after the model is initialized, we start the actual training in which we concatenate all the requested feature vectors as well as the expected outcome into the RNN model repeatedly for the whole training set until 100 test iterations have passed. The number of actual epochs trained in each iteration is configurable. In

Table 4. Used event logs and their relevant statistics

Event log	# Cases	# Activities	#*Events*	# Attributes	# Unique values
BPIC12[a]	13087	24	262200	1	3
BPIC13, incidents[b]	7554	13	65533	8	2890
BPIC14[c]	46616	39	466737	1	242
BPIC17[d]	31509	26	1202267	4	164
BPIC18[e]	43809	41	2514266	5	360

[a]https://doi.org/10.4121/uuid:3926db30-f712-4394-aebc-75976070e91f
[b]https://doi.org/10.4121/uuid:500573e6-accc-4b0c-9576-aa5468b10cee
[c]https://doi.org/10.4121/uuid:c3e5d162-0cfd-4bb0-bd82-af5268819c35
[d]https://doi.org/10.4121/uuid:5f3067df-f10b-45da-b98b-86ae4c7a310b
[e]https://doi.org/10.4121/uuid:3301445f-95e8-4ff0-98a4-901f1f204972

our experiments, the total number of epochs was set to be 10. After every test iteration, the model is validated against the validation set. To improve validation performance, if the size of the validation set is larger than a separately specified limit (10000), a random sample of the whole validation set is used. These test results, including additional status and timing-related information, are written into resulting test result CSV file. If the prediction accuracy of the model against the validation set is found to be better than the accuracy of any of the models found thus far, then the network state is stored for that model. Finally, after all the training, the model having the best validation test accuracy is picked as the actual result of the model training.

5.2 Testing

In the testing phase, the third subset of cross-validation folding is tested against the model built in the previous step. After initializing the event log following similar steps as in the training phase, the model is asked for a prediction for each input vector built from the test data. To prevent running out of memory and to ensure tests are not taking an exceedingly long time to run, we limited the number of final test traces to 100000 traces and used random sampling when needed. The prediction result accuracy, as well as other required statistics, are written to the resulting CSV file.

6 Test Setup

We performed our tests using several different data sets. Some details of the used data sets can be found in the Table 4. The table lists the number of cases, activities, events and event attributes for each event log. *# Unique values* column shows the sum of all the unique attribute values for each of the selected attributes.

Table 5. Included event attributes

Event log	Attribute names
BPIC12	lifecycle:transition
BPIC13, incidents	impact
	lifecycle:transition
	org:group
	org:resource
	organization country
	organization involved
	product
	resource country
BPIC14	Assignment Group
BPIC17	Action
	EventOrigin
	lifecycle:transition
	org:resource
BPIC18	activity
	doctype
	note
	org:resource
	subprocess

The criteria for selecting or not selecting an event attribute in a model were based on the maximum usage of any unique value that the attribute has in the event log. If a value of an attribute was used in more than 4% of all the events in the event log, then that attribute was included in the test. Besides, we did not select any attributes that had just one unique attribute value. Names of all the selected event attributes are listed in the Table 5.

For each dataset, we performed the next activity prediction where we wanted to predict the next activity of any ongoing case. This was accomplished by splitting every input case into possibly multiple *virtual* cases depending on the number of events the case had. If the length of the case was shorter than 4, the whole case was ignored. If the length was equal or higher, then a separate *virtual* case was created for all prefixes at least of length 4. Thus, for a case of length 6, 3 cases were created: One with length 4, one with 5 and one with 6. For all these prefixes, the next activity label was used as the expected outcome. For the full-length case, the expected outcome was a special *finished*-token.

7   Experiments

For experiments, we have used the same system that we used already in our previous work [9]. The system had Windows 10 operating system and its hardware
consisted of 3.5 GHz Intel Core i5-6600K CPU with 32 GB of main memory and
NVIDIA GeForce GTX 960 GPU having 4 GB of memory. Out of those 4 GB, we
reserved 3 GB for the tests. The testing framework was built on the test system
using the Python programming language. The actual recurrent neural networks
were built using Lasagne[2] library that works on top of Theano[3]. Theano was
configured to use GPU via CUDA for expression evaluation.

We used one layer GRU [2] as the RNN type. Adam [11] was used as gradient
descent optimizer with parameters of $beta_1 = 0.9$ and $beta_2 = 0.999$. 256 was
used as the hidden dimension size as well as the mini-batch size and 0.01 as
the learning rate. Even though it is quite probable that more accurate results
could have been achieved by selecting, e.g., different hidden dimension sizes and
learning rates depending on the size of the input vectors, we decided to use
the fixed values. This decision was made in order to make the interpretation of
the test results easier by minimizing the number of variables affecting the test
results.

We performed next activity predictions using all the four combinations of
features, five data sets and three different maximum cluster counts: 20, 40, and
80 clusters. The results of these runs are shown in Table 6. In the table, *Features*
-column shows the used set of features. *S.rate* shows the achieved prediction
success rate. *In.v.s.* shows the size of the input vector. This column can be used
to give some kind of indication on the memory usage of using that configuration.
Finally, *Tra.t.* and *Pred.t.* columns tell us the time required for performing the
training and the prediction for all the cases in the test dataset. In both cases,
this time includes the time for setting up the neural network, clusterings and
preparing the dataset from JSON format. Sample standard deviation has been
included in both *S.rate* and *Tra.t* in parentheses to indicate how spread out the
measurements are within all the three test runs. Each row in the table represents
three cross-validation runs with a unique combination of dataset and feature that
was tested. Rows having the best prediction accuracy within a dataset are shown
using bold font. *None* -feature represents the case in which there were no event
attribute information at all in the input vector, *ClustN* represents a test with
one-hot encoded cluster labels of event attributes clustered into maximum of N
clusters, *Raw* represents having all one-hot encoded attribute values individually
in the input vector, and finally, *BothN* represents having both one-hot encoded
attribute values and one-hot encoded cluster labels in the input vector.

We also aggregated some of these results over all the datasets using the maximum cluster size of 80 clusters. Figure 1 shows the average success rates of different event attribute encoding techniques over all the tested datasets. Figure 2
shows the average input vector lengths. Figures 3 and 4 shows the averaged train-

---

[2] https://lasagne.readthedocs.io/.
[3] http://deeplearning.net/software/theano/.

Table 6. Statistics of next activity prediction using different sets of input features

Dataset	Features	S.rate (σ)	In.v.s.	Tra.t. (σ)	Pred.t.
BPIC12	None	85.8% (0.3%)	25.7	489.0 s (7.0 s)	35.1 s
	Clust20	86.0% (0.4%)	30.0	500.6 s (2.5 s)	31.6 s
	Clust40	85.8% (0.3%)	30.0	499.7 s (1.3 s)	31.9 s
	Clust80	86.2% (0.1%)	30.0	502.1 s (2.3 s)	7.5 s
	Raw	85.9% (0.3%)	29	504.3 s (0.5 s)	38.9 s
	Both20	86.0% (0.2%)	33	515.3 s (2.6 s)	40.4 s
	Both40	86.0% (0.4%)	33	517.7 s (3.6 s)	40.4 s
	Both80	**86.3% (0.1%)**	**33**	**518.2 s (4.0 s)**	**40.7 s**
BPIC13	None	62.9% (0.3%)	13.7	165.6 s (21.2 s)	3.5 s
	Clust20	66.8% (0.3%)	34.7	188.0 s (22.4 s)	4.7 s
	Clust40	67.2% (0.7%)	54.7	214.8 s (3.1 s)	5.4 s
	Clust80	67.0% (0.6%)	94.7	258.4 s (4.7 s)	6.0 s
	Raw	68.2% (1.1%)	2353.7	2611.7 s (44.7 s)	74.8 s
	Both20	**69.1% (0.6%)**	**2359.3**	**2464.6 s (309.0 s)**	**94.4 s**
	Both40	68.9% (0.5%)	2395.7	2687.1 s (227.3 s)	106.6 s
	Both80	68.4% (0.7%)	2429.3	2821.8 s (33.5 s)	194.3 s
BPIC14	None	37.8% (1.5%)	40.3	488.1 s (5.3 s)	36.1 s
	Clust20	39.9% (0.5%)	61.7	523.3 s (3.5 s)	40.4 s
	Clust40	40.0% (0.3%)	80.3	553.5 s (3.8 s)	43.6 s
	Clust80	40.2% (0.1%)	84.7	556.8 s (10.5 s)	43.6 s
	Raw	39.7% (1.4%)	272.0	825.7 s (2.8 s)	68.0 s
	Both20	40.6% (0.6%)	292.3	907.1 s (7.5 s)	78.6 s
	Both40	**40.6% (0.6%)**	**309.3**	**943.3 s (10.6 s)**	**82.0 s**
	Both80	37.3% (4.2%)	305.0	935.1 s (26.9 s)	156.7 s
BPIC17	None	86.4% (0.4%)	27.7	518.7 s (2.8 s)	107.7 s
	Clust20	**90.8% (0.3%)**	**48.7**	**556.3 s (3.7 s)**	**132.4 s**
	Clust40	90.2% (1.4%)	68.3	637.5 s (58.3 s)	143.9 s
	Clust80	90.2% (0.4%)	108.7	647.3 s (3.7 s)	142.8 s
	Raw	89.9% (0.5%)	190	816.4 s (5.2 s)	164.9 s
	Both20	89.9% (0.5%)	211.0	867.8 s (3.5 s)	188.0 s
	Both40	90.2% (0.3%)	230.3	910.9 s (19.3 s)	193.7 s
	Both80	89.6% (0.6%)	271.3	986.5 (4.4 s)	197.7 s
BPIC18	None	71.3% (9.3%)	43	516.0 s (9.5 s)	197.0 s
	Clust20	79.0% (0.9%)	64.0	588.7 s (13.7 s)	268.7 s
	Clust40	**79.9% (0.2%)**	**84.0**	**628.4 s (2.8 s)**	**286.1 s**
	Clust80	79.5% (0.1%)	124.0	701.3 s (7.4 s)	306.9 s
	Raw	79.3% (0.4%)	349.7	1173.7 s (83.1 s)	381.2 s
	Both20	79.7% (0.5%)	377.7	1213.1 s (48.1 s)	463.2 s
	Both40	79.9% (0.5%)	401.0	1301.9 s (82.9 s)	540.3 s
	Both80	79.3% (0.5%)	425.7	1405.4 s (87.2 s)	619.9 s

ing and prediction times respectively. Finally, Fig. 5 shows the average success rates of different event attribute encoding techniques over all the tested datasets in the case where the maximum cluster count was set to be 80.

Next, we measured the statistical significance of the results. First, we measured whether we could reject the null hypothesis: "the best success rate results could have been achieved by using *Raw* features". By performing a one-tailed $t - test$, while assuming equal variances, for the results, we find out that this hypothesis can not be completely rejected in any of the test datasets. However,

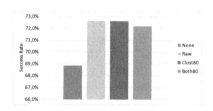

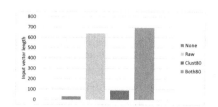

Fig. 1. Average prediction success rate over all the datasets

Fig. 2. Average length of the input vector over all the datasets

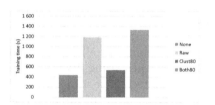

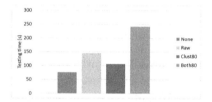

Fig. 3. Average training time over all the datasets

Fig. 4. Average prediction time over all the datasets

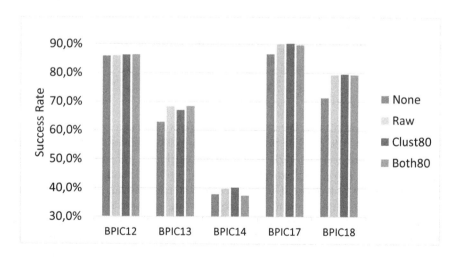

Fig. 5. Average prediction success rate over all the datasets separately

when assessing the null hypothesis: "we can achieve better success rates without taking event attributes into account at all than by taking them into account as clustered attribute values", we can reject it in all the datasets. Similarly, we can easily reject null hypothesis: "training a model using *Raw* features can be as fast as using the most accurate tested clustered features" in all the other cases, except in BPIC12, where the used input vectors were essentially identical due to the small number of unique attribute values in that model.

Based on all of the test results and statistical significance analysis, we can see that having event attribute values included improved the prediction accuracy over not having them included at all in all datasets. The effect ranged from 0.5% in BPIC12 model to 8.5% in BPIC18. As shown in Fig. 1, very similar success rates were achieved using *ClustN* features as with *Raw*. However, model training and the actual prediction can be performed faster using *ClustN* approaches than either *Raw* or *BothN*. This effect is the most prominently visible in BPIC13 results, where, due to the model having a large amount of unique attribute values, the size of the input vector is almost 68 times bigger and the training time almost 14 times longer using *Raw* feature than *Clust20*. At the same time, the accuracy is still better than not having event attributes at all (about 3.9% better) and only slightly worse (about 1.4%) than when using *Raw* feature. This indicates that clustering can be a really powerful technique for minimizing the time required for training especially when there are a lot of unique event attribute values in the used event log. Even when using the maximum cluster count of 20, prediction results will be either not affected or improved with a relatively small impact on the training and prediction time. Input vector size and memory usage are affected by clustered features in a similar fashion with the training and testing time: The more there are unique attribute values in the event log, the greater the difference between the input vector sizes needed in *ClustN* and *Raw* features.

In all the datasets, the best prediction accuracy is always achieved either by using only clustering or by using both clustering and raw attributes at the same time.

7.1   Threats to Validity

As threats to the validity of the results in this paper, it is clear that there are a lot of variables involved. As the initial set of parameter values, we used parameters that were found good enough in our earlier work and did some improvement attempts based on the results we got. It is most probable that the set of parameters we used were not optimal ones in each test run. We also did not test all the parameter combinations and the ones we did, we tested often only once, even though there was some randomness involved, e.g., selecting the initial cluster centers in the XMeans algorithm. However, we think that since we tested the results in several different datasets using a 3-fold cross-validation technique, our results can be used at least as a baseline for further studies. All the results generated by the test runs, as well as all the source data and the test framework itself, are available in support materials[4].

Also, we did not test with datasets having a huge number of event attribute values, the maximum amount tested being 2890. However, it can be seen that since the size of the input vectors is completely user-configurable when performing event attribute clustering, the user him/herself can easily set limits to the input vector length which should take the burden off from the RNN and move the burden to the clustering algorithms, which are usually more efficient in handling lots of features and feature values. When evaluating the results of the performed tests and comparing them with other similar works, it should be taken into account that data sampling was used in several phases of the testing process.

8   Conclusions

Clustering can be applied to attribute values to improve the accuracy of predictions performed on running cases. In four of the five experimented data sets, having event attribute clusters encoded into the input vectors outperforms having the actual attribute values in the input vector. Also, due to raw attribute values having direct effect to input vector lengths, the training and prediction time will be directly affected by the number of unique event attribute values. Clustering does not have this problem: The number of elements reserved in the input vector for clustered event attribute values can be adjusted freely. The memory usage is directly affected by the length of the input vector. In the tested cases, the number of clusters to use to get the best prediction accuracy seemed to depend very much on the used datasets, when the tested cluster sizes were 20, 40 and 80. In some cases, having more clusters improved the performance, whereas, in others, it did not have any significant impact, or even made the accuracy worse. We also found out that in some cases, having attribute cluster indicators in the input vectors improved the prediction even if the input vectors also included all the actual attribute values.

As future work, it would be interesting to test this clustering approach also with other machine learning model types such as more traditional random forest and gradient boosting machines. Similarly, it could be interesting to first filter out some of the most rarely occurring attribute values before clustering the values. This could potentially reduce the amount of noise added to the clustered data and make it easier for the clustering algorithm to not be affected by noisy data. Another idea that we leave for future study is whether it would be a good idea to first perform some kind of a feature selection algorithm such as influence analysis [12], recursive feature elimination [7] or mRMR [3] to find the attribute values that correlate the most with the prediction results and have those attribute values added into the input vectors as raw one-hot encoded attribute values in addition to having the one-hot encoded cluster labels. More work is also required to understand exactly what properties of the event log

---

[4] https://github.com/mhinkka/articles.

affect the optimal number of clusters to use. Finally, more study is required to understand whether a similar clustering approach performed for event attributes in this work could be applicable also for encoding case attributes.

Acknowledgments. We want to thank QPR Software Plc for funding our research. Financial support of Academy of Finland project 313469 is acknowledged.

References

1. Bernard, G., Andritsos, P.: Accurate and transparent path prediction using process mining. In: Welzer, T., Eder, J., Podgorelec, V., Kamišalić Latifić, A. (eds.) ADBIS 2019. LNCS, vol. 11695, pp. 235–250. Springer, Cham (2019). https://doi.org/10.1007/978-3-030-28730-6_15

2. Cho, K., van Merrienboer, B., Bahdanau, D., Bengio, Y.: On the properties of neural machine translation: encoder-decoder approaches. In: Wu, D., Carputat, M., Carreras, X., Vecchi, E.M. (eds.) Proceedings of SSST@EMNLP 2014, Eighth Workshop on Syntax, Semantics and Structure in Statistical Translation, Doha, Qatar, 25 October 2014, pp. 103–111. Association for Computational Linguistics (2014)

3. Ding, C.H.Q., Peng, H.: Minimum redundancy feature selection from microarray gene expression data. J. Bioinf. Comput. Biol. **3**(2), 185–206 (2005)

4. Evermann, J., Rehse, J., Fettke, P.: Predicting process behaviour using deep learning. Decis. Support Syst. **100**, 129–140 (2017)

5. Francescomarino, C.D., Dumas, M., Maggi, F.M., Teinemaa, I.: Clustering-based predictive process monitoring. CoRR, abs/1506.01428 (2015)

6. Francescomarino, C.D., Ghidini, C., Maggi, F.M., Milani, F.: Predictive process monitoring methods: which one suits me best? In: Weske et al. [22], pp. 462–479 (2018)

7. Granitto, P.M., Furlanello, C., Biasioli, F., Gasperi, F.: Recursive feature elimination with random forest for PTR-MS analysis of agroindustrial products. Chemometr. Intell. Lab. Syst. **83**(2), 83–90 (2006)

8. Hartigan, J.A., Wong, M.A.: Algorithm AS 136: a k-means clustering algorithm. J. R. Stat. Soc. Ser. C (Appl. Stat.) **28**(1), 100–108 (1979)

9. Hinkka, M., Lehto, T., Heljanko, K., Jung, A.: Structural feature selection for event logs. In: Teniente, E., Weidlich, M. (eds.) BPM 2017. LNBIP, vol. 308, pp. 20–35. Springer, Cham (2018). https://doi.org/10.1007/978-3-319-74030-0_2

10. Hinkka, M., Lehto, T., Heljanko, K., Jung, A.: Classifying process instances using recurrent neural networks. In: Daniel, F., Sheng, Q.Z., Motahari, H. (eds.) BPM 2018. LNBIP, vol. 342, pp. 313–324. Springer, Cham (2019). https://doi.org/10.1007/978-3-030-11641-5_25

11. Kingma, D.P., Ba, J.: Adam: a method for stochastic optimization. CoRR, abs/1412.6980 (2014)

12. Lehto, T., Hinkka, M., Hollmén, J.: Focusing business improvements using process mining based influence analysis. In: La Rosa, M., Loos, P., Pastor, O. (eds.) BPM 2016. LNBIP, vol. 260, pp. 177–192. Springer, Cham (2016). https://doi.org/10.1007/978-3-319-45468-9_11

13. Navarin, N., Vincenzi, B., Polato, M., Sperduti, A.: LSTM networks for data-aware remaining time prediction of business process instances. In: 2017 IEEE Symposium Series on Computational Intelligence, SSCI 2017, Honolulu, HI, USA, 27 November–1 December 2017, pp. 1–7. IEEE (2017)

14. Nolle, T., Seeliger, A., Mühlhäuser, M.: BINet: multivariate business process anomaly detection using deep learning. In: Weske et al. [22], pp. 271–287 (2018)
15. Pelleg, D., Moore, A.W.: X-means: extending k-means with efficient estimation of the number of clusters. In: Langley, P. (ed.) Proceedings of the Seventeenth International Conference on Machine Learning (ICML 2000), Stanford University, Stanford, CA, USA, 29 June–2 July 2000, pp. 727–734. Morgan Kaufmann, Burlington (2000)
16. Rousseeuw, P.J.: Silhouettes: a graphical aid to the interpretation and validation of cluster analysis. J. Comput. Appl. Math. **20**, 53–65 (1987)
17. Schönig, S., Jasinski, R., Ackermann, L., Jablonski, S.: Deep learning process prediction with discrete and continuous data features. In: Damiani, E., Spanoudakis, G., Maciaszek, L.A. (eds.) Proceedings of the 13th International Conference on Evaluation of Novel Approaches to Software Engineering, ENASE 2018, Funchal, Madeira, Portugal, 23–24 March 2018. SciTePress, Setubal, pp. 314–319. SciTePress, Setubal (2018)
18. Tax, N., Teinemaa, I., van Zelst, S.J.: An interdisciplinary comparison of sequence modeling methods for next-element prediction. CoRR, abs/1811.00062 (2018)
19. Tax, N., Verenich, I., La Rosa, M., Dumas, M.: Predictive business process monitoring with LSTM neural networks. In: Dubois, E., Pohl, K. (eds.) CAiSE 2017. LNCS, vol. 10253, pp. 477–492. Springer, Cham (2017). https://doi.org/10.1007/978-3-319-59536-8_30
20. van der Aalst, W.M.P.: Process Mining - Discovery, Conformance and Enhancement of Business Processes. Springer, Berlin (2011). https://doi.org/10.1007/978-3-642-19345-3
21. Verenich, I., Dumas, M., Rosa, M.L., Maggi, F.M., Chasovskyi, D., Rozumnyi, A.: Tell me what's ahead? predicting remaining activity sequences of business process instances, June 2016
22. Weske, M., Montali, M., Weber, I., vom Brocke, J. (eds.): BPM 2018. LNCS, vol. 11080. Springer, Cham (2018). https://doi.org/10.1007/978-3-319-98648-7

General Model for Tracking Manufacturing Products Using Graph Databases

Jorge Martinez-Gil[1(✉)], Reinhard Stumptner[2], Christian Lettner[1],
Mario Pichler[1], Salma Mahmoud[1], Patrick Praher[1],
and Bernhard Freudenthaler[1]

[1] Software Competence Center Hagenberg GmbH, Softwarepark 21,
4232 Hagenberg, Austria
jorge.martinez-gil@scch.at
[2] FAW GmbH, Softwarepark 35, 4232 Hagenberg, Austria

Abstract. One of the major problems in the manufacturing industry consists of the fact that, when manufacturing a product, many parts from different lots are supplied and mixed to a certain degree during an indeterminate number of stages, what makes it very difficult to trace each of these parts from its origin to its presence in a final product. In order to overcome this limitation, we have worked towards the design of a general solution aiming to improve the traceability of the products from several manufacturers. This solution is based on the exploitation of graph databases which allows us to significantly reduce response times compared to traditional relational systems.

Keywords: Data engineering · Graph databases · Knowledge
Graphs · Manufacturing

1 Introduction

Quality related errors in manufacturing create a lot of problems for the industrial sector mainly because they lead to a great waste of resources in terms of time, money and effort spent to identify and solve them [17]. For this reason, researchers and practitioners aim to find novel solutions capable of tracking and analyzing manufacturing products in an appropriate and easy to use manner [14]. There are already several approaches belonging to different manufacturing domains: additive manufacturing [16], toy manufacturing [4], electrical equipment [12], or fabrication of cylindrical markers [11].

This work is an extended version of: Jorge Martinez-Gil, Reinhard Stumptner, Christian Lettner, Mario Pichler, and Werner Fragner. *Design and implementation of a graph-based solution for tracking manufacturing products*. In New Trends in Databases and Information Systems. ADBIS Workshops 2019. Communications in Computer and Information Science, volume 1064, pages 417–423. Springer Cham, 2019.

P. Ceravolo et al. (Eds.): SIMPDA 2018/2019, LNBIP 379, pp. 86–100, 2020.
https://doi.org/10.1007/978-3-030-46633-6_5

In this context, one of the challenges of modern manufacturing is that a final product usually consists of different components which themselves can also consist of different components, and so forth. At the lowest level, there is raw material (e.g. steel coils, sheet metal, steel rolls, etc.) which nature is also very relevant to the quality of the final product. In this way, tracking and connecting all the data of the different manufacturing stages is crucial to finding the causes of quality-related errors in the final product.

In case the components and raw materials have one-to-one or one-to-many relationships to the final products, the tracking process is quite straightforward and has already been implemented satisfactorily in some of the existing solutions. However, manufacturing products can also be made up of parts that are in lots, and lots can be combined to make other lots. Also, many different lots can be combined into a new one that can be part of many other ones. This means that some manufacturers have to deal with a problem involving many-to-many relationships with blurred relationships among the single parts.

One of the most important problems here is that providing these relationships is very important for both the operators and the quality engineers, so they can drill down from the final product to the assembled components and ultimately to the used raw materials and can identify causes to problems that are not obvious at the first glance.

Unfortunately, the existing solutions based on relational databases are not very useful for the people who are in charge of examining these lots. On the one hand, the existing solutions have bad response times when dealing with this problem, and on the other hand, there are no meaningful probability values available. So the quality engineers must invest a lot of time to analyze all possibly related data and cannot just focus on the relevant data.

To alleviate this problem, we have tried to look for a solution so that it can be possible to track all items from different lots that were used in final products. Our proposed solution is based on the exploitation of graph databases. The major advantage of such approach, concerning the existing ones, is that it allows for informed queries, i.e. queries that can lead to early termination if nodes with no compatible outgoing relations are found. As a result, we have got a software system that presents lower execution time for most of the use cases concerning the tracking of items.

In our previous work [19], we presented a specific approach for the design and solution of a tailored solution for a manufacturing company located in Upper Austria. In the present work, our major contribution here is an extension of our previous work to build a general solution for appropriately tracking the manufacturing products that have different kinds of dependencies (evolution, distribution, and so on). Our solution is intended to outperform traditional systems based on relational databases in the specific context of tracking defective items in lots of manufacturing products. Besides, to illustrate our proposal, we include a complete use case that shows some of the functionality that can be derived from a solution of this kind.

The rest of this work is structured in the following way: Sect. 2 presents the state-of-the-art concerning the current graph-based solutions for the manufacturing industry. Section 3 describes the design and implementation of our solution and a use case whereby our solution outperforms the traditional tracking systems. Section 4 discusses the lessons learned from this research. Finally, in Sect. 5, we remark the conclusions that can be extracted from this work as well as the possible future lines of research.

2 State-of-the-art

In spite of the great need for tracking solutions in manufacturing environments, it seems that most of the quality assurance processes that require controlling and supervising the whole production chain to timely detect human errors and defective materials need to be further automatized. The major reason is that little attention has been paid to this problem due to technical limitations, and as result, there are not too many solutions in the manufacturing domain, but just a few works have been proposed to date [9,20,24,25].

We can go even a step further beyond to see that the problem can be aggravated, even in the case of passing all quality assurance controls. The reason is that products can be rejected by end-users or other manufacturers if unknown problems appear. Therefore, the capability to track each part of a lot from its origin is of vital importance for the manufacturing industry.

In this context, it is necessary to remark that existing manufacturing systems are far from being trivial since the data models they work with often consist of many data types and data sources that are in no relation to each other. Due to this, modeling and optimization, as well as process analysis, represent often a hard task [26].

One of the major limiting factors for the solution to this problem is that traditional relational database systems (i.e. the systems that have so far been used mostly in the industry) are unable to model this specific problem effectively. Therefore, we have focused our research on graph databases [23]. The idea behind graph databases is their capability to store data in nodes and edges versus tables, as found in relational databases [3]. Each node represents an entity, and each edge represents a relationship between two nodes [5]. This way of modeling the problem is much more natural and is in line with a problem arising from dependencies such as this.

It is generally assumed that graph databases have some key advantages over relational databases in this context. The reason is that unlike relational databases, graph databases are designed to store interconnected data what makes it easier to work with these data by not forcing intermediate indexing at every time, and also making it easier to facilitate the evolution of the data that we are working with.

In the literature, recurrent mention is made of some of the advantages of graph databases concerning traditional relational models. It has been possible to identify the three major advantages of graph databases in comparison with traditional relational databases to tackle this problem:

(1) The first advantage when dealing with a graph is that as opposed to the relational world, foreign key relationships are not relations in the sense of edges of a graph.

(2) The second important advantage of graph databases in relation to relational databases is that, when referring to the latter ones, it is not possible to assign properties or labels to relationships. It is possible to give them a name in the database, but it is not possible to visualize them (e.g. derivations, transformations, etc.). When, in fact, in graph databases is the natural way to model data.

(3) Last, but not least, traditional systems based on the exploitation of relational databases are not able to scale as well as graph databases when dispatching relationship-like queries [2].

Some of these advantages are making graph databases gaining popularity among big industrial players. Moreover, their application domain is very broad [1]. In fact, many organizations are already using databases of this kind for detecting fraud in monetary transactions, providing product and service recommendations, documenting use cases and lessons learned in a wide range of domains, managing access control in restricted places, network monitoring to identify potential risks and hazards, and so on. Moreover, if we focus strictly on the manufacturing industry, graph-based solutions have already been proposed in forecasting and recommendation [22].

3   General Model for Tracking Manufacturing Products Using Graph Databases

To illustrate the problem that we are facing with an example, let us think of a situation where a finished part of a product is rejected by the customer because of a number of quality errors. In that case, it is quite common that the manufacturer of the finished part must make a statement within 24 h if this error can be restricted to the single rejected product or if a greater amount of parts is affected. In case a greater amount of parts is affected, then the exactly affected lots must be reported to the customer.

As it is not difficult to imagine, a situation of this kind happens very often, and the worst thing is not that, but that it is really expensive in terms of effort, time, money and brand image. So it should be tried by all means to minimize its impact as far as possible.

To do that, the first task of the manufacturer is to try to find the cause of the quality error. Therefore it must be possible to analyze the captured data of the finished part but also the captured data of all assembled components and raw materials. In case the cause of the quality error lies in a specific component or raw material lot, the manufacturer must find all finished parts that contain this lot. This means that finding the right affected lots in a short period of time can decide whether the manufacturer must recall millions of finished parts or just a few. This means that if we can identify the affected lots quickly, it is not necessary to recall all delivered lots.

It is possible to think a situation whereby to perform this analytic task, the manufacturer needs to search back and forth across all the data of the finished parts. If a relational database is used, this means that many queries and their corresponding responses would need to be combined. In our approach, the solution is much more intuitive since it is in general possible to easily write queries capable of running over the data in any direction as we will see later in this paper. The capability to discover and see the connections between different parts of a product allows a human operator to effectively perform this tracking.

3.1 Notation

Let L_m^t be a lot of parts produced on machine m at time step t. Every machine m has a buffer b_m where lots to be processed at this machine are poured into, i.e. they are getting blurred. Then, the relation $usage : (L_m^{t2}, L_n^{t1})$ defines that at time slice $t2$ the lot L_m^{t1} has been poured into the buffer of machine m for processing. So beginning at time slice $t2$ parts of lot L_n^{t1} are installed with a certain probability into parts of lot L_m^{t2}.

The relation $pred : \{(L_m^{t2}, L_m^{t1})\}$ defines that lot L_m^{t1} is produced before lot L_m^{t2} on machine m and the buffer b_m of machine m was not empty when the production of L_m^{t2} started. Lots that are delivered by other suppliers, i.e. raw materials, are treated the same way. They will be assigned a virtual production machine number and time slice which uniquely identifies the batch number from the supplier.

Using the same mathematical notation, simple examples for one-to-one and the one-to-many relationship between lots without blurring are depicted in Fig. 1 and 2. An example with a many-to-many relationship that includes blurring of lots is given in Fig. 3. By following the edges of the graph, the lots that are built in other lots can be determined easily, i.e. for lot L_3^4 parts of the lots $\{L_1^1, L_1^2, L_2^2, L_2^3\}$ may be included, or for lot L_3^5 parts of the lots $\{L_1^2, L_1^3, L_2^3, L_2^4\}$ may be included.

The missing relation $pred$ between L_3^4 and L_3^5 indicates that the buffer of machine 3 was empty before the production of lot L_3^5 started so no blurring of lots could have occurred. The distance between lots can be used as a basis to determine the probability a certain part of a lot is used in another lot. A more precise determination of probabilities would also require to consider buffer levels during manufacturing but that is beyond the scope of this work.

3.2 Implementation

We have implemented a prototypical software solution to we can see how a given tracking system could use the graph database OrientDB[1] in order to implement common operations. We have chosen OrientDB, since it is a multimodal NoSQL (which stands for not only SQL) database that combines properties of document-oriented and graph databases [7]. It allows users to define graph structures using

---

[1] https://orientdb.com/.

Fig. 1. One-to-one relationship between lots. This is the simplest relationship that we can find in the manufacturing industry. In principle, a solution for dealing with this type of relationship is quite simple and can be implemented efficiently in a wide range of database systems, including those databases making use of the traditional model.

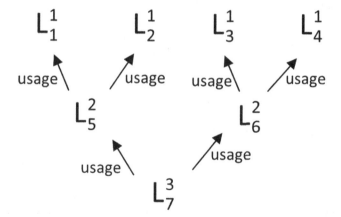

Fig. 2. One-to-many relationship between lots. It is a fairly common type of relationship in the manufacturing industry. The content of one batch is distributed or transformed in turn into other batches. This type of relationship can also be implemented efficiently in most current database systems, including relational systems.

concepts for nodes and edges but also allows us to append complex data to nodes in the form of documents. Nodes and edges can have attributes (e.g. edge weight or similar).

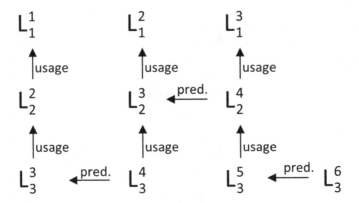

Fig. 3. Many-to-many relationship between lots i.e. blurred lots. It is a complex, yet a fairly common, type of relationship in the manufacturing industry. It takes place when many lots evolve or are distributed among other lots making their trace very difficult to follow. We hypothesize that only graph database systems can model and implement a solution efficiently. This type of relationship represents the central problem around which our research work revolves.

Moreover, inheritable classes can be defined for the nodes and edges which can be extended flexibly. The query language is an adapted form of SQL and it is very intuitive and easy to use. In fact, the queries can be easy expressed through a user interface as represented in Fig. 4.

It is important to note that the smallest unit that can be loaded and saved from the database is a record. OrientDB distinguishes between four types of records: A record can be a document, a RecordBytes (BLOB), a node (Vertex) or an edge (Edge). In our specific case, working with Vertex and Edge data types is enough. But as future work, we can envision that it would be useful to use the data type document to offer explicit support when dealing with situations that have been already before. Therefore, choosing OrientDB also represents an efficient and scalable alternative, which is why an OrientDB-based implementation has been set up.

The rationale behind the election of this solution is that, compared to implementations based on relational database systems, using a graph database leads to an efficient and scalable solution in which the problem at hand can be modeled easily [13]. Maybe the most clear example is the action of traversing graphs. The fact is that traversing graphs modeled in relational database systems would require to write nested and recursive queries that are difficult to maintain and provide bad comparable performance.

The following SQL source code shows an brief example of how to model the problem using a traditional relational database approach to illustrate our viewpoint.

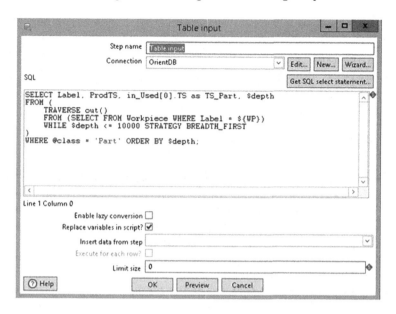

Fig. 4. OrientDB graphical interface that allows users to design and launch queries related to the distribution and/or evolution of the different lots through the manufacturing process.

```
1   create table workpiece (
2     ts int,
3     wp_name varchar(50)
4   )
5
6   create table partlot (
7     ts int,
8     lot_name varchar(50)
9     )
10
11  insert into workpiece values (1,'WP1')
12  insert into workpiece values (2,'WP2')
13  insert into workpiece values (3,'WP3')
14  insert into workpiece values (4,'WP4')
15  insert into workpiece values (5,'WP5')
16  insert into workpiece values (6,'WP6')
17
18  insert into partlot values (1,'L1')
19  insert into partlot values (2,'L2')
20  insert into partlot values (5,'L3')
21
22  select * from workpiece
23  select * from partlot
```

```
24
25  select
26    l.lot_name ,
27    ( select COUNT(*)
28      from workpiece p
29      where p.ts  >= l.ts
30        and p.ts < 5 -- query for workpiece WP6
31    ) distance
32  from partlot l
```

This source code shows how workpieces and their relations (in the form of belonging to a lot) can be modeled. We need a table for workpieces and a different one for lots, Then by means of insertions we can store the corresponding data. Finally, it is necessary to design a query to get the results. However, when using our approach we can appreciate several advantages. In this way we can represent the problem in a very natural way and proceed to the implementation of a solution that is both fast and efficient.

It is important to remark that traversing a graph is the act of visiting the nodes in the graph. For graph databases, traversing to nodes via their relationships can be compared to the join operations on the relational database tables.

The great advantage of this solution is that the operation for traversing a graph is much faster than the traditional joins from the relational databases world. The reason is that when querying the database with a traversal, the model only considers the data that is needed without taking into account any kind of grouping operations on the entire data, as it happens in traditional relational databases.

3.3 Use Cases

Based on the implementation that has been described above, such as the graph-based approach is considered to have a positive impact on the daily operations of the manufacturing industry. In particular, the depth calculation, i.e. the distance between lots, could be used as a basis for calculating the dependency path of a part in a product.

In order to illustrate our approach, we show here some examples of queries that our system can support. The following code shows how to calculate the shortest path between nodes #33:27050 and #29:2667 regardless of edge direction. Note that the unwind directive is useful while performing the aggregation of the nodes in the path.

```
1  SELECT expand(path) FROM (
2    SELECT shortestPath(#33:27050, #29:2667)
3           AS path UNWIND path
4  );
```

Figure 5 shows the resulting graph from the calculation of the shortest path between nodes #33:27050 and #29:2667 regardless of edge direction.

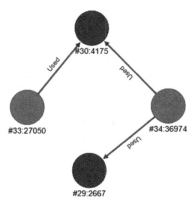

Fig. 5. Result of calculating the shortest path between nodes #33:27050 and #29:2667 regardless of edge direction.

OrientDB solution already implements the so-called search with the so-called BREADTH FIRST, which returns the real depth in the graph. It is trivial to see that the greater the depth, the less likely it is that each lot has been incorporated into the end product. The following code results in Fig. 6, and it shows us how to traverse the graph in the direction of the edge direction from node #33:27050 to a depth of 10.

```
1  TRAVERSE out ()
2  FROM #33:27050 WHILE depth < 10 STRATEGY BREADTH_FIRST;
```

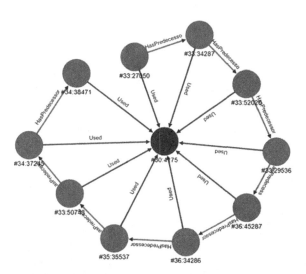

Fig. 6. Result of traversing the graph from node #33:27050 to a depth of 10.

An example of SQL-like query to get the shortest path between lot 33 : 27050 and 29 : 2667 can be easily written as:

```
1  SELECT expand(path) FROM (
2  SELECT shortestPath(#33:27050, #29:2667, 'OUT')
3              AS path UNWIND path);
```

The result of this query is depicted in Fig. 7. The number of edges between the lots gives a basic notion of probability that parts of lot 29 : 2667 are built-in parts of lot 33 : 27050. As we have seen before, the higher the number of nodes visited by the query, the lower the probability that each of the specific nodes is affected by the error or failure that the operator or quality engineer are looking for.

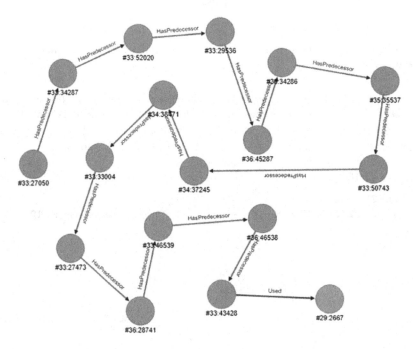

Fig. 7. Result of calculating the shortest path between the nodes #33:27050 and #29:2667.

4   Discussion

During the last years, the ever-increasing technical literature concerning graph-based research clearly shows us that one of the current main challenges of computer science consists of finding proper ways to model the knowledge generated in a specific domain [18].

With this regard, Knowledge Graphs [21] have gained some popularity recently. In this respect, we believe that the manufacturing domain fits very well.

This opinion is shared with other researchers who have already been working in this direction to be able to explore the knowledge graph by the industry [15].

In this context, some software solutions being able to facilitate the handling of product-related production for manufacturing enterprises in the industry sector are becoming more popular [10]. Some of the most popular models to represent domain knowledge are based on some kind of the so-called knowledge graphs [8]. Knowledge graphs are intended to represent entities and relationships between entities within a particular universe of discourse. The advantage of using this kind of knowledge representation is that it is easy to understand for both humans and computers at the same time.

The origin of knowledge graphs is the combination of the knowledge bases with the inference engines, what is also referred to as Knowledge-Based Systems in the literature. One of the first approaches that one could think about would be a system that represents knowledge with uncertainty using a set of rules to the that they are given a certainty factor. However, these types of systems based on rules are not very robust, so they have been progressively replaced by another type of more efficient system, being at present the Bayesian networks the most used way of representing and inferring interesting knowledge with uncertainty currently.

Using this idea of knowledge-based systems, Google launched the so-called Google Knowledge Graph [6] several years ago. This graph seeks to have a universal domain, representing all existing entities and relationships between them, without being subject to a single context. Having such a domain broad has great complexity, and is incomplete. It is complex to introduce a new entity and determine what other entities it relates to and what type of relationship unites them.

This task of generating new entities is not currently automated but is the users themselves who introduce new entities to the network and determine what other entities relate to each other and how. It is expected that within the next years, this technology will be developed and it is possible that some of its foundations can be applied in the manufacturing domain.

5   Conclusions and Future Work

In this work, we have presented the design of a general solution for tracking each component that comprises manufacturing products through diverse stages of the production chain. Our solution has been modeled using graph databases as opposed to most existing solutions that use relational databases. In this way, our approach provides an improved level of both transparency and traceability, since we think that a graph is the natural way to model a problem involving dependencies of this kind.

Transparency is given by the fact that it is for a human operator to see what actions have been performed during the manufacturing process. Traceability is given by the fact that it is possible to monitor the whole development process followed by the manufacturer. In addition, these two properties are assumed to facilitate the analysis of all manufacturing products as well as the capability to look for final products that could be affected by some specific problems. In this way, our approach presents more efficient modeling and querying mechanisms than traditional approaches based on relational databases.

As future research work, we envision that graph databases hold a lot of unrealized potential in the next years as companies will be moving towards approaches being able to better data analysis and exploration. More specifically, we think that there is a number of pending challenges. For example, it is important to further investigate whether it is possible to document past errors, to facilitate the task of discovering future errors. To do this, it is necessary to investigate whether it is possible to associate documentation with certain error patterns, which have to happen recurrently during the manufacturing process in the past. We also believe that the use of Knowledge Graphs can play a determining role, since these graphs allow not only to model the problem in a natural way, but also to reason with the data we work with.

Acknowledgements. This work has been supported by the project **AutoDetect** (Project No. 862019; Innovative Upper Austria 2020 (call Digitalization)) as well as the Austrian Ministry for Transport, Innovation and Technology, the Federal Ministry of Science, Research and Economy, and the Province of Upper Austria in the frame of the COMET center SCCH.

References

1. El Abri, M.: Probabilistic relational models learning from graph databases. (Apprentissage des modèles probabilistes relationnels à partir des bases de données graphe). Ph.D. thesis, University of Nantes, France (2018)
2. Angles, R., Arenas, M., Barceló, P., Hogan, A., Reutter, J.L., Vrgoc, D.: Foundations of modern query languages for graph databases. ACM Comput. Surv. **50**(5), 68:1–68:40 (2017)
3. Balaghan, P.: An exploration of graph algorithms and graph databases. Ph.D. thesis, University of Hull, Kingston upon Hull, UK (2019)
4. Cao, Y., Li, W., Song, W., Chaovalitwongse, W.A.: Collaborative material and production tracking in toy manufacturing. In: Proceedings of the 2013 IEEE 17th International Conference on Computer Supported Cooperative Work in Design (CSCWD), Whistler, BC, Canada, 27–29 June 2013, pp. 645–650 (2013)
5. Cheng, J., Ke, Y., Ng, W.: Efficient query processing on graph databases. ACM Trans. Database Syst. **34**(1), 2:1–2:48 (2009)
6. Färber, M., Bartscherer, F., Menne, C., Rettinger, A.: Linked data quality of DBpedia, Freebase, OpenCyc, Wikidata, and YAGO. Semant. Web **9**(1), 77–129 (2018)
7. Fernandes, D., Bernardino, J.: Graph databases comparison: Allegrograph, ArangoDB, InfiniteGraph, Neo4J, and OrientDB. In: Proceedings of the 7th International Conference on Data Science, Technology and Applications, DATA 2018, Porto, Portugal, 26–28 July 2018, pp. 373–380 (2018)
8. Galkin, M., Auer, S., Scerri, S.: Enterprise knowledge graphs: a backbone of linked enterprise data. In: 2016 IEEE/WIC/ACM International Conference on Web Intelligence (WI 2016), Omaha, NE, USA, 13–16 October 2016, pp. 497–502 (2016)

9. Ghazel, M., Toguyéni, A., Bigand, M.: A semi-formal approach to build the functional graph of an automated production system for supervision purposes. Int. J. Comput. Integr. Manuf. **19**(3), 234–247 (2006)
10. He, L., Jiang, P.: Manufacturing knowledge graph: a connectivism to answer production problems query with knowledge reuse. IEEE Access **7**, 101231–101244 (2019)
11. Hülß, J.-P., Müller, B., Pustka, D., Willneff, J., Zürl, K.: Tracking of manufacturing tools with cylindrical markers. In: The 18th ACM Symposium on Virtual Reality Software and Technology (VRST 2012), Toronto, ON, Canada, 10–12 December 2012, pp. 161–168 (2012)
12. Ivanov, V., Brojboiu, M.D., Ivanov, S.: Applications of the graph theory for optimization in manufacturing environment of the electrical equipments. In: 28th European Conference on Modelling and Simulation (ECMS 2014), Brescia, Italy, 27–30 May 2014, pp. 153–158 (2014)
13. Jouili, S., Vansteenberghe, V.: An empirical comparison of graph databases. In: International Conference on Social Computing (SocialCom 2013) SocialCom/PAS-SAT/BigData/EconCom/BioMedCom 2013, Washington, DC, USA, 8–14 September 2013, pp. 708–715 (2013)
14. Kammler, F., Hagen, S., Brinker, J., Thomas, O.: Leveraging the value of data-driven service systems in manufacturing: a graph-based approach. In: 27th European Conference on Information Systems - Information Systems for a Sharing Society (ECIS 2019), Stockholm and Uppsala, Sweden, 8–14 June 2019 (2019)
15. Kudryavtsev, D., Gavrilova, T., Leshcheva, I.A., Begler, A., Kubelskiy, M., Tushkanova, O.: Mind mapping and spreadsheets in collaborative design of manufacturing assembly units' knowledge graphs. In: Joint Proceedings of the BIR 2018 Short Papers, Workshops and Doctoral Consortium co-located with 17th International Conference Perspectives in Business Informatics Research (BIR 2018), Stockholm, Sweden, 24–26 September 2018, pp. 82–93 (2018)
16. Livesu, M., Cabiddu, D., Attene, M.: slice2mesh: a meshing tool for the simulation of additive manufacturing processes. Comput. Graph. **80**, 73–84 (2019)
17. Lou, K.-R., Wang, L.: Optimal lot-sizing policy for a manufacturer with defective items in a supply chain with up-stream and down-stream trade credits. Comput. Ind. Eng. **66**(4), 1125–1130 (2013)
18. Martinez-Gil, J.: Automated knowledge base management: a survey. Comput. Sci. Rev. **18**, 1–9 (2015)
19. Martinez-Gil, J., Stumpner, R., Lettner, C., Pichler, M., Fragner, W.: Design and implementation of a graph-based solution for tracking manufacturing products. In: Welzer, T., et al. (eds.) ADBIS 2019. CCIS, vol. 1064, pp. 417–423. Springer, Cham (2019). https://doi.org/10.1007/978-3-030-30278-8_41
20. Minoufekr, M., Driate, A., Plapper, P.W.: An IoT framework for assembly tracking and scheduling in manufacturing SME. In: Proceedings of the 16th International Conference on Informatics in Control, Automation and Robotics (ICINCO 2019) - Volume 2, Prague, Czech Republic, 29–31 July 2019, pp. 585–594 (2019)
21. Noy, N.F., Gao, Y., Jain, A., Narayanan, A., Patterson, A., Taylor, J.: Industry-scale knowledge graphs: lessons and challenges. Commun. ACM **62**(8), 36–43 (2019)
22. Ringsquandl, M., Lamparter, S., Lepratti, R.: Graph-based predictions and recommendations in flexible manufacturing systems. In: IECON 2016–42nd Annual Conference of the IEEE Industrial Electronics Society, Florence, Italy, 23–26 October 2016, pp. 6937–6942 (2016)

23. Robinson, I., Webber, J., Eifrem, E.: Graph Databases. O'Reilly Media Inc., Newton (2013)
24. Singh, M., Khan, I.A., Grover, S.: Selection of manufacturing process using graph theoretic approach. Int. J. Syst. Assur. Eng. Manag. **2**(4), 301–311 (2011). https://doi.org/10.1007/s13198-012-0083-z
25. Wang, H., Yang, J., Ceglarek, D.J.: A graph-based data structure for assembly dimensional variation control at a preliminary phase of product design. Int. J. Comput. Integr. Manuf. **22**(10), 948–961 (2009)
26. Weise, J., Benkhardt, S., Mostaghim, S.: A survey on graph-based systems in manufacturing processes. In: IEEE Symposium Series on Computational Intelligence (SSCI 2018), Bangalore, India, 18–21 November 2018, pp. 112–119 (2018)

Supporting Confidentiality in Process Mining Using Abstraction and Encryption

Majid Rafiei[1]([✉])[ID], Leopold von Waldthausen[2][ID],
and Wil M. P. van der Aalst[1][ID]

[1] Chair of Process and Data Science, RWTH Aachen University, Aachen, Germany
majid.rafiei@pads.rwth-aachen.de
[2] Yale University, New Haven, USA

Abstract. Process mining aims to bridge the gap between data science and process science by providing a variety of powerful data-driven analyses techniques on the basis of event data. These techniques encompass automatically discovering process models, detecting and predicting bottlenecks, and finding process deviations. In process mining, event data containing the full breadth of resource information allows for performance analysis and discovering social networks. On the other hand, event data are often highly sensitive, and when the data contain private information, privacy issues arise. Surprisingly, there has currently been little research toward security methods and encryption techniques for process mining. Therefore, in this paper, using *abstraction*, we propose an approach that allows us to hide confidential information in a controlled manner while ensuring that the desired process mining results can still be obtained. We show how our approach can support confidentiality while discovering control-flow and social networks. A connector method is applied as a technique for storing associations between events securely. We evaluate our approach by applying it on real-life event logs.

Keywords: Responsible process mining · Confidentiality · Process discovery · Directly follows graph · Social network analysis

1 Introduction

Data science is changing the way we do business, socialize, conduct research, and govern society. Data are collected on anything, at any time, and in any place. Therefore, it is not surprising that many people are concerned about the responsible use of data. The Responsible Data Science (RDS) [8] initiative focuses on four main questions: (1) How to avoid unfair conclusions even if they are true?, (2) How to answer questions with a guaranteed level of accuracy?, (3) How to answer questions without revealing secrets?, and (4) How to clarify answers such that they become indisputable? This paper focuses on the confidentiality problem (third question) when applying process mining to event data.

© IFIP International Federation for Information Processing 2020
Published by Springer Nature Switzerland AG 2020
P. Ceravolo et al. (Eds.): SIMPDA 2018/2019, LNBIP 379, pp. 101–123, 2020.
https://doi.org/10.1007/978-3-030-46633-6_6

Process mining uses event data to provide novel insights into actual processes [2]. There are many activities and techniques in the field of process mining. However, the three basic types of process mining are; process discovery [1], conformance checking [2], and process re-engineering (enhancement) [7]. Also, four perspectives are considered to analyze the event data including; *control-flow*, *organizational*, *case*, and *time* perspective [2]. In this paper, we focus on *control-flow* and *organizational* perspective side by side. A simple definition for process discovery is learning process models from event logs. In fact, a discovery technique takes an event log and produces process model without using additional information [5]. A social network is a social structure which shows relations among social actors (individuals or organizations) [30]. When event data contain information about resources, not only can it be used to thoroughly analyze bottlenecks, but also it turns to a valuable data to derive social networks among resources, involved in the process. Since such event data contain highly sensitive information about the organization and the people involved in the process, confidentiality is a major concern. Note that by confidentiality in process mining, we aim to deal with two important issues; (1) protecting the sensitive data belonging to the organization, (2) protecting the private information about the individuals.

As we show in this paper, *confidentiality in process mining cannot be achieved by merely encrypting all data.* Since people need to use and see process mining results, the challenge is to retain as little information as possible while still being able to have the same desired result. Here, the desired results are process models and social networks. The discovered models (networks) based on the anonymized event data should be identical to the results obtained from the original event data (assuming proper authorizations).

In this paper, we propose an approach to deal with confidentiality in process mining which is based on *abstractions*. Moreover, we present the *connector* method by which the individual traces of a process stay anonymous, yet, at the same time, process models and social networks are discoverable. The proposed framework allows us to derive the same results from secure event logs when compared to the results from original event logs, while unauthorized persons cannot access confidential information. In addition, this framework can provide a secure solution for process mining when processes are cross-organizational.

The remainder of this paper is organized as follows. Section 2 outlines related work and the problem background. In Sect. 3, we clarify process mining, social network discovery, and cryptography as preliminaries. In Sect. 4, the problem is explained in detail. Our approach is introduced in Sect. 5. In Sect. 6 the approach is evaluated, and Sect. 7 concludes the paper.

2   Related Work

In data science, social networks, and information systems, confidentiality has been a topic of interest in the last decade. In computer science, privacy-preserving algorithms and methods in differential privacy are most applicable to confidentiality in process mining. In sequential pattern mining, the field of data science which arguably close to process mining, there has been work on preserving privacy in settings with distributed databases [15] or in cross-organizational settings [31]. Also, privacy-preservation in social networks is a well-researched topic, and most of the research in this field aims to protect the privacy of the individuals involved in a given social network [17]. However, here, we focus on the confidentiality issues arising when initially discovering social networks from event logs that comprise lots of sensitive private data about the individuals.

Although there have been a lot of breakthroughs in the field of process mining ranging from data preprocessing [28] and process discovery [22] to performance analysis [18] and prediction [25], the research field confidentiality and privacy has received relatively little attention. This is despite the fact that already the Process Mining Manifesto [6] points out that privacy concerns are important to be addressed. In the following, we introduce some research regarding *Responsible Process Mining (RPM)* and few publications which focused specifically on confidentiality issues, in the *control-flow* perspective or during *process discovery*.

The topic of Responsible Process Mining (RPM) [3] has been put forward by several authors thereby raising concerns related to fairness, accuracy, confidentiality, and transparency. In [29], a method for securing event logs to be able to do process discovery by Alpha algorithm has been proposed. In [12], a possible approach toward a solution, allowing the outsourcing of process mining while ensuring the confidentiality of dataset and processes, has been presented. In [20], the authors has used a cross-organizational process discovery setting, where public process model fragments are shared as safe intermediates. In [23], the aim is to provide an overview of privacy challenges when process mining is used in human-centered industrial environments. In [27], the authors introduce a framework for ensuring confidentiality in process mining which is utilized and extended in this paper. In [14], a privacy model is proposed for privacy-aware process discovery. In [26], the organizational perspective in process mining is taken into account, and the aim is to provide a privacy-preserving method for role mining, which can be used for generalizing *resources* as individuals in event data. It is also worth noting that process mining can be used for security analyses, e.g., in [10], process mining is used for security auditing.

3 Background

In this section, we briefly present the main concepts and refer the readers to relevant literature for more detailed explanations.

3.1 Process Mining

In the following, we introduce some basic concepts of process mining to which we will refer in this paper.

Events are the smallest data unit in process mining and occur when an activity in a process is executed. Events comprise of multiple attributes including; *Case ID, Timestamp, Activity, Resource*, etc. In Table 1, each row indicates an event. In the remainder of this paper, we will refer to the activities and resources of Table 1 with their abbreviations, e.g., "R" stands for "Register".

A *trace* is a sequence of events and represents how a process is executed in one instance, e.g., in Table 1, case 1 is first registered, then documents are verified, and vacancies are checked. Finally, a decision is made for the case.

An *event log* is a collection of sequences of events which are used as the input of process mining algorithms. Event data are widely available in current information systems [6].

As you can see in Table 1, a "Timestamp" identifies the moment in time at which an event has taken place, and a "Case ID" is what all events in a trace have in common so that they can be identified as part of that process instance. Event logs can also include additional attributes for the events they record. There are two main attribute types that fall under this category. "Event Attributes" which are specific to an event, and "Case Attributes" which are ones that stay the same throughout an entire trace.

Table 1. Sample event log (each row represents an event).

Case ID	Timestamp	Activity	Resource	Cost
1	01-01-2018:08.00	Register (R)	Frank (F)	1000
2	01-01-2018:10.00	Register (R)	Frank (F)	1000
3	01-01-2018:12.10	Register (R)	Joey (J)	1000
3	01-01-2018:13.00	Verify-Documents (V)	Monica (M)	50
1	01-01-2018:13.55	Verify-Documents (V)	Paolo (P)	50
1	01-01-2018:14.57	Check-Vacancies (C)	Frank (F)	100
2	01-01-2018:15.20	Check-Vacancies (C)	Paolo (P)	100
4	01-01-2018:15.22	Register (R)	Joey (J)	1000
2	01-01-2018:16.00	Verify-Documents (V)	Frank (F)	50
2	01-01-2018:16.10	Decision (D)	Alex (A)	500
5	01-01-2018:16.30	Register (R)	Joey (J)	1000
4	01-01-2018:16.55	Check-Vacancies (C)	Monica (M)	100
1	01-01-2018:17.57	Decision (D)	Alex (A)	500
3	01-01-2018:18.20	Check-Vacancies (C)	Joey (J)	50
3	01-01-2018:19.00	Decision (D)	Alex (A)	500
4	01-01-2018:19.20	Verify-Documents (V)	Joey (J)	50
5	01-01-2018:20.00	Special-Case (S)	Katy (K)	800
5	01-01-2018:20.10	Decision (D)	Katy (K)	500
4	01-01-2018:20.55	Decision (D)	Alex (A)	500

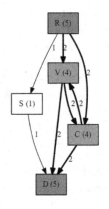

Fig. 1. The DFG resulting from event log Table 1

A Directly Follows Graph (DFG) is a graph where the nodes represent activities and the arcs represent causalities. Activities "a" and "b" are connected by an arrow when "a" is frequently followed by "b". The weights of the arrows denote the frequency of the relation [19]. Most commercial process mining tools use DFGs. Unlike more advanced process discovery techniques (e.g., implemented in ProM), DFGs cannot express concurrency. Figure 1 shows the DFG resulting from the event log Table 1.

3.2   Discovering Social Networks

There are different methods for discovering social networks from event logs including those based on *causality*, *joint activities*, *joint cases*, etc. [9]. Here,

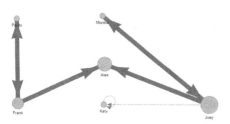

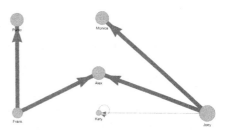

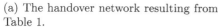

(a) The handover network resulting from Table 1.

(b) The real-handover network resulting from Table 1 for the real causal dependency threshold 0.1.

Fig. 2. The networks based on causality for the event log Table 1

we, however, focus purely on the metrics based on causality. These metrics monitor for individual cases how work moves from resource to resource. E.g., there is a *handover* relation from individual i to individual j, if there are two subsequent activities where the first is performed by i and the second is performed by j. This relation furthermore becomes a *real-handover* if casual dependency between both activities exists. Note that in this case the directly follows relations between resources are not enough and the real casual dependencies are required. Dependency measure (Eq. 1) can be used to realize whether there is a real casual dependency between two activities (a and b) or not, while a threshold is set as the minimum required value [2]. In Eq. 1, $|a>_L b|$ shows how frequent a is followed by b:

$$|a \Rightarrow_L b| = \begin{cases} \frac{|a>_L b| - |b>_L a|}{|a>_L b| - |b>_L a| + 1} & if\ a \neq b \\ \frac{|a>_L b|}{|a>_L b| + 1} & if\ a = b \end{cases} \tag{1}$$

When observing handovers, indirect succession may also be considered. E.g., based on the event log of Table 1, there is a non-real *handover* relation between "Frank" and "Alex" with the depth three. It is non-real due to there is no real casual dependency between all the corresponding activities. Figure 2 shows the networks based on causality having been obtained from event log Table 1.

3.3 Cryptography

Cryptography or cryptology is about constructing and analyzing protocols that prevent third parties or the public from reading private messages [11].

A cryptosystem is a suite of cryptographic algorithms needed to implement a particular security service, most commonly for achieving confidentiality [16]. There are different kinds of cryptosystems. In this paper, we use the following ones.

– *Symmetric Cryptosystem:* The same secret key is used to encrypt and decrypt a message. Data manipulation in symmetric systems is faster than asymmetric systems as they generally use shorter key lengths. Advanced Encryption Standard (AES) is a symmetric encryption algorithm [13].

- *Asymmetric Cryptosystem:* Asymmetric systems use a public key to encrypt a message and a private key to decrypt it or vice versa. The use of asymmetric systems enhances the security of communication. Rivest-Shamir-Adleman (RSA) is an asymmetric encryption algorithm.
- *Deterministic Cryptosystem:* A deterministic cryptosystem is a cryptosystem which always produces the same ciphertext for a given plaintext and key, even over separate executions of the encryption algorithm.
- *Probabilistic Cryptosystem:* A probabilistic cryptosystem, other than the deterministic cryptosystem, is a cryptosystem which uses randomness when encrypting so that when the same plaintext is encrypted several times, it will produce different ciphertexts.
- *Homomorphic Cryptosystem:* A homomorphic cryptosystem allows computation on ciphertext, e.g., Paillier is a partially homomorphic cryptosystem [24].

4   Problem Definition (Attack Analysis)

To illustrate the challenge of confidentiality in process mining, we start this section with an example. Consider Table 2, describing a totally encrypted event log, belonging to a hospital conducting surgeries. Since we need to preserve difference to find a sequence of activities for each case, discovering process model, and other analyses like social network discovery, "Case ID", "Activity", and "Resource" are encrypted based on a deterministic encryption method. Numerical data (i.e., "Timestamp" and "Cost") are encrypted by a homomorphic encryption method to preserve the ability of basic mathematical computations on the encrypted data. Now suppose that we have background knowledge about surgeons and the approximate cost of different types of surgeries. The question arises whether parts of the log can now be deanonymized.

Owning to the fact that "Cost" is encrypted by a homomorphic encryption method, the maximum value for the "Cost" is the real maximum cost and based on background knowledge we know that e.g., the most expensive event in the hospital was the brain surgery by "Dr. Jone", on "01/09/2018 at 12:00", and the patient name is "Judy". Since "Case ID", "Activity", and "Resource" are encrypted by a deterministic encryption method, we can replace all these encrypted values with the corresponding plain values. Consequently, encrypted data could be made visible without requiring decryption. This example demonstrates that even given completely encrypted event logs small fraction of contextual knowledge can leads to data leakage.

Given domain knowledge, several analyses could be done to identify individuals or extract some sensitive information from an encrypted event log. In the following, we explain couple of them.

- *Exploring the Length of Traces:* One can find the longest/shortest trace, and the related background knowledge can be exploited to realize the actual activities and the related case(s).

– *Frequency Mining:* One can find the most or the less frequent traces and the related background knowledge can be utilized to identify the corresponding case(s) and the actual activities.

These are just some examples demonstrate that encryption alone is not a solution. For example, [21] shows that mobility traces are easily identifiable after encryption. Any approach which is based on solely encrypting the whole event log will furthermore have the following weaknesses:

– *Encrypted Results:* Since results are encrypted, the data analyst is not able to interpret the results. E.g., as data analyst we want to know which paths are the most frequent after "Registration" activity; how can one perform this analysis when the activities are not plain? The only solution is decrypting the results.
– *Impossibility of Accuracy Evaluation:* How can we make sure that a result of the encrypted event log is the same as the result of the plain event log? Again, decryption would be required.

Generally, and as explored by [12], using cryptography is a resource consuming activity, and decryption is even much more resource consuming than encryption. The weaknesses demonstrate that encryption methods should be used wisely and one needs to evaluate closely where they are beneficiary and where unavoidable to provide confidentiality.

Here, we assume that background knowledge could be any contextual knowledge about traces which can result in a *case disclosure* including; frequency of traces, length of traces, exact/approximate time related to the cases, etc. Note that this background knowledge is assumed where unauthorized people can access the anonymized data. For example, given domain knowledge regarding frequency of traces one can guess the actual sequence of activities and possible case(s) (e.g., politicians, celebrities, etc) for the traces which are too rare. Consequently, individuals or minority group of people and their private information would be revealed. Therefore, the *case disclosure* is a crucial type of data leakage which should be prevented.

Table 2. A totally encrypted event log.

Case ID	Activity	Resource	Timestamp
rt!@45	kl56ˆ*	lo09(kl	3125
rt!@45	bn,.ˆq	lo09(kl	3256
)@!1yt	kl56ˆ*	lo09(kl	4879
)@!1yt	bvS(op	/.,ldf	5214
)@!1yt	jhg!676	nb][,b]	6231
erˆ7*	kl56ˆ*	lo09(kl	6534
erˆ7*	2ws34S	v,[]df	7230

5   Approach

Figure 3 illustrates a framework to provide a solution for confidentiality when the desired result is a model. This framework has been inspired by [5], where *abstractions* are introduced as intermediate results for relating models and logs. Here, *abstractions* are directly follows matrix of activities (A-DFM) and directly follows matrix of resources (R-DFM). Figure 4 shows the A-DFM and R-DFM resulting from event log Table 1. A-DFM is considered as the abstraction for relating logs and process models, and R-DFM together with A-DFM are considered as the abstraction for relating logs and social networks which are based on causality. As can be seen in Fig. 3 three different environments and two confidentiality solutions are presented.

- *Forbidden Environment:* In this environment, the actual information system runs that needs to use the real data. The real event logs (EL) produced by this environment contain a lot of valuable confidential information and except some authorized persons no one can access this data.
- *Internal Environment:* This environment is just accessible by authorized stakeholders. A data analyst can be considered as an authorized stakeholder who can access the internal event logs. Event logs in this environment are partially secure, selected results produced in this environment (e.g., a process model) are the same as the results produced in the forbidden environment, and data analyst is able to interpret the results without decryption.
- *External Environment:* In this environment, unauthorized external persons can access the data. Such environments may be used to provide the computing infrastructure dealing with large data sets (e.g., a cloud solution). Event logs in this environment are supposed to be entirely secure, and the results are encrypted. Whenever a data analyst wants to interpret results, they need to be decrypted and converted to an internal version. Furthermore, results from the external environment do not need to be exactly the same as the results from the internal environment, but, the same interpretations need to be provided.

Table 3 shows a summary of our assumptions with respect to the internal and external environments. Note that in the forbidden environment, the main assumption is that only few highly trusted persons can access the data. Therefore, there is no need to employ confidentiality solutions. As described in Sect. 4, contextual knowledge regarding traces is assumed as background knowledge. As can be seen in Fig. 3, the desired results, which are process models (PM) and social networks (SN), can be obtained in each environment. The original event log (EL) is converted to the partially secure event log in the internal environment (EL') and then to the entirely secure event log in the external environment (EL'') by the internal confidentiality solution (ICS) and the external confidentiality solution (ECS) respectively. *Abstractions*, which are intermediate results, are used for proving accuracy. It should be taken into account that since *abstractions* are considered as the outputs of the very last phase before the final results

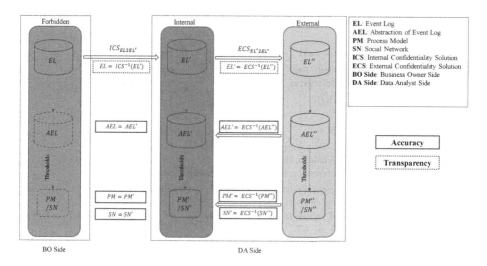

Fig. 3. The proposed framework for confidentiality in process mining.

(only thresholds are required to be applied), when they are equal, the final results would be the same. In addition, *transparency* is provided by the reverse operation of the internal confidentiality solution (ICS^{-1}) and the reverse operation of the external confidentiality solution (ECS^{-1}). In the following, we explain our ICS and ECS in detail.

5.1   Internal Confidentiality Solution (ICS)

For ICS we combine several methods and introduce the connector method. Figure 5 gives an overview of the anonymization procedure.

Filtering and Modifying the Input. The first step to effective anonymization is preparing the data input. To filter the input, simple limits for frequencies can be set, and during loading an event log all traces that do not reach the minimal frequencies are not transferred to the EL'.

	C	D	R	S	V
C	0	2	0	0	2
D	0	0	0	0	0
R	2	0	0	1	2
S	0	1	0	0	0
V	2	2	0	0	0

	Alex	Frank	Joey	Katy	Monica	Paolo
Alex	0	0	0	0	0	0
Frank	2	0	0	0	0	2
Joey	2	0	0	1	2	0
Katy	0	0	0	1	0	0
Monica	0	0	2	0	0	0
Paolo	0	2	0	0	0	0

(a) The A-DFM resulting from Table 1.   (b) The R-DFM resulting from Table 1.

Fig. 4. The *abstractions* from Table 1.

Table 3. The general assumptions based on the environments

	Internal	External
Who has access to the data?	Employees Internal Data Analysts	External Data Analysts Anyone else
Trust to the ones who have data access	High	Low
Background knowledge	Broad	Limited
What data should be kept secure?	Direct individual/organization sensitive data which is not necessary for the desired result	Direct/indirect individual/organization sensitive data
Desired results	-Social network based on causality from EL' -Process model based on DFG from EL'	-Social network based on causality from EL'' -Process model based on DFG from EL''

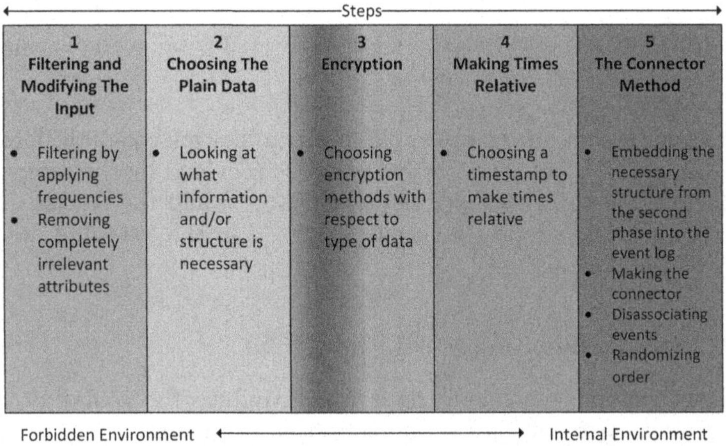

Fig. 5. The internal confidentiality solution.

Choosing the Plain Data. As mentioned, we need to produce interpretable results. Hence, some parts of event log remain as plain text in the internal version of the secure event log (EL'). We should decide what information and/or structure is strictly necessary for the desired analysis. Based on our considered abstractions (A-DFM and R-DFM), the only information necessary are directly follows relations between activities/resources.

Encryption. Here there are two important choices. The first choice is which columns of the event log should be encrypted. Second, we need to decide which algorithms should be used. For the internal environment, since we want to keep the capability of applying basic mathematical computations on the encrypted values, we use Paillier for numeric attributes (i.e., "Cost"), and AES-128 with

Table 4. The first 10 rows of Table 1 after encryption and making times relative

Case ID	Timestamp	Activity	Resource	Cost
1	00-00-0000:08.00	Register (R)	Frank (F)	0820315
2	00-00-0000:10.00	Register (R)	Frank (F)	0820315
3	00-00-0000:12.10	Register (R)	Joey (J)	0820315
3	00-00-0000:13.00	Verify-Documents (V)	Monica (M)	0650210
1	00-00-0000:13.55	Verify-Documents (V)	Paolo (P)	0650210
1	00-00-0000:14.57	Check-Vacancies (C)	Frank (F)	0650900
2	00-00-0000:15.20	Check-Vacancies (C)	Paolo (P)	0650900
4	00-00-0000:15.22	Register (R)	Joey (J)	0820315
2	00-00-0000:16.00	Verify-Documents (V)	Frank (F)	0650210
2	00-00-0000:16.10	Decision (D)	Alex (A)	0710155

only ASCII characters as the key is used for other attributes. Note that the encrypted values shown in the paper are not necessarily the real outputs of the encryption methods (they are just unintelligible text).

Making Times Relative. Times need to be modified because keeping the exact epoch time of an event can allow one to identify it. The naive approach, of setting the starting time of every trace to 0, would make it impossible to replay events and reconstruct the original log. Thus, we select another time that all events are made relative to. This time can be kept secure along with the keys for decryption. Table 4 shows the first 10 rows of our sample log after encrypting cost and making times relative to the "01-01-2018:00.00".

Table 5. Adding previous activities/resources and previous IDs.

Case ID	Timestamp	Activity	Prev. Activity	Resource	Prev. Resource	Cost	ID	Prev. ID
1	00-00-0000:08.00	R	START	Frank (F)	START	0820315	31	00
2	00-00-0000:10.00	R	START	Frank (F)	START	0820315	32	00
3	00-00-0000:12.10	R	START	Joey (J)	START	0820315	38	00
3	00-00-0000:13.00	V	R	Monica (M)	Joey (J)	0650210	41	38
1	00-00-0000:13.55	V	R	Paolo (P)	Frank (F)	0650210	55	31
1	00-00-0000:14.57	C	V	Frank (F)	Paolo (P)	0650900	09	55
2	00-00-0000:15.20	C	R	Paolo (P)	Frank (F)	0650900	86	32
4	00-00-0000:15.22	R	START	Joey (J)	START	0820315	47	00
2	00-00-0000:16.00	V	C	Frank (F)	Paolo (P)	0650210	75	86
2	00-00-0000:16.10	D	V	Alex (A)	Frank (F)	0710155	56	75

Table 6. The event log after adding the connector column

Case ID	Timestamp	Activity	Prev. Activity	Resource	Prev. Resource	Cost	ID	Prev. ID	Connector
1	00-00-0000:08.00	R	START	Frank (F)	START	0820315	31	00	1<@sadd21?
2	00-00-0000:10.00	R	START	Frank (F)	START	0820315	32	00	!s*f*+dsf3
3	00-00-0000:12.10	R	START	Joey (J)	START	0820315	38	00	ça/ds23"w'
3	00-00-0000:13.00	V	R	Monica (M)	Joey (J)	0650210	41	38	.,m;lo,mh
1	00-00-0000:13.55	V	R	Paolo (P)	Frank (F)	0650210	55	31	;l4;l,'kjh
1	00-00-0000:14.57	C	V	Frank (F)	Paolo (P)	0650900	09	55	*';k!kjm."
2	00-00-0000:15.20	C	R	Paolo (P)	Frank (F)	0650900	86	32	l:mj/.m @p
4	00-00-0000:15.22	R	START	Joey (J)	START	0820315	47	00	;k;lm.lå@,
2	00-00-0000:16.00	V	C	Frank (F)	Paolo (P)	0650210	75	86	=ó@k;d/f.m
2	00-00-0000:16.10	D	V	Alex (A)	Frank (F)	0710155	56	75	';,lk.;hj!

The Connector Method. Using the connector method we embed the structure, which can be used for extracting directly follows relations, into EL'. Also, the connector method helps us to reconstruct the full original event logs when keys and relative values are given. In the first step, the previous activity ("Prev. Activity") and the previous resource ("Prev. Resource") columns are added in order to identify which arcs can be directly connected.

In the second step, we find a way to securely save the information contained in the "Case ID", without allowing it to link the events. This can be done by giving each row a random ID ("ID") and a previous ID ("Prev. ID"). These uniquely identify the following event in a trace because the IDs are not generic like activity names. The ID for start activities is always a number of zeros. Table 5 shows the log after adding "Prev. Activity", "Prev. Resource", "ID", and "Prev. ID".

In the third step, regarding the fact that these columns contain the same information previously found in the "Case ID", they must be hidden and secured. This can be done by concatenating the "ID" and "Prev. ID" of each row and

Table 7. The output event log after applying ICS

Timestamp	Activity	Prev. Activity	Resource	Prev. Resource	Cost	Connector
08.00	R	START	Frank (F)	START	0820315	1<@sadd21?
01.02	C	V	Frank (F)	Paolo (P)	0650900	!s*f*+dsf3
10.00	R	START	Frank (F)	START	0820315	ça/ds23"w'
15.22	R	START	Joey (J)	START	0820315	.,m;lo,mh
00.50	V	R	Monica (M)	Joey (J)	0650210	;l4;l,'kjh
00.40	V	C	Frank (F)	Paolo (P)	0650210	*';k!kjm."
12.10	R	START	Joey (J)	START	0820315	l:mj/.m @p
05.20	C	R	Paolo (P)	Frank (F)	0650900	;k;lm.lå@,
05.55	V	R	Paolo (P)	Frank (F)	0650210	=ó@k;d/f.m
00.10	D	V	Alex (A)	Frank (F)	0710155	';,lk.;hj!

encrypting those using AES. Due to the nature of AES, neither orders nor sizes of the IDs remain inferable. The concatenation can be done in any style, in this example, we however simply concatenate "ID" and "Prev. ID",e.g., connector of the first row would be "3100". To retain the "ID" and "Prev. ID" one simply needs to decrypt the "Connector" column and cut the resulting number in two equal parts. This method requires that every time the two IDs differ by a factor 10 a zero must be added to guarantee equal length. Table 6 shows the log after concatenating the ID columns and encrypting them as a connector.

In the final step, we use the "Case ID" to anonymize the "Time tamp". The "Time tamp" attribute of events which have the same "Case ID" is made relative to the preceded one. The exception is the first event of each trace which remains unchanged. This allows the complete calculation of all durations of the arcs in a directly follows graph but makes it complicated to identify events based on the epoch times they occurred at. After creating the relative times, we are free to delete the "Case ID" and disarray the order of all rows, ending up with an unconnected log in Table 7.

Table 7 is internally secure event log (EL'), which can be used by a data analyst to create a A-DFM and a R-DFM. It is trivial to see that if process/social network discovery could have been done on the plain event log (EL), AEL would be identical to AEL' (i.e., both are the same A-DFM/R-DFM) and the final desired results would be the same. Note that when the desired result is a process model, resource related information ("Resource" and "Prev. Resource" columns) can be removed from Table 7. Moreover, when the desired result is a handover network, activity related information ("Activity" and "Prev. Activity") can be removed, since the real causal dependencies do not need to be taken into account.

Comparing Table 7 and the original log, one can see that there is no answer for the following questions in EL' anymore: (1) *Who was responsible for doing an activity for case c?* (2) *What is the sequence of activities which has been done for case c?* (3) *How long did it take to process case c?* (4) *What is the cost of activity a which has been done by resource r for case c?* (5) *What is the the length of case c?* (6) *What is the the frequency of case c?*, and many other questions related to the cases.

It is also worth noting that since we assume that the data in the internal environment can be accessed by the internal trustworthy people who already know the organizational structure, the plain resources are not considered as a privacy issue. In fact, EL' is a partially secure version of event log in such a way that it contains the minimum level of information, which a data analyst might need to reach the result. Although ICS does not preserve the standard format of the event log which is used by the current process discovery techniques, the intermediate input it provides can be used by the current tools. In the External Confidentiality Solution (ECS), we need to avoid any form of data leakage and privacy risks based on the assumed background knowledge.

Table 8. The event log after encrypting activities and resources

Timestamp	Activity	Prev. Activity	Resource	Prev. Resource	Cost	Connector
08.00	AgeIRL	1wBo2I	908G2F	1wBo2I	0820315	1<@sadd21?
01.02	5rYd7h	v42jbE	908G2F	9iYoqT	0650900	!s*f*+dsf3
10.00	AgeIRL	1wBo2I	908G2F	1wBo2I	0820315	ça/ds23"w'
15.22	AgeIRL	1wBo2I	RjjZyw	1wBo2I	0820315	.,m;lo,mh
00.50	v42jbE	AgeIRL	eBzosT	RjjZyw	0650210	;l4;l,'kjh
00.40	v42jbE	5rYd7h	908G2F	9iYoqT	0650210	*';k!kjm."
12.10	AgeIRL	1wBo2I	RjjZyw	1wBo2I	0820315	l:mj/.m @p
05.20	5rYd7h	AgeIRL	9iYoqT	908G2F	0650900	;k;lm.lå@,
05.55	v42jbE	AgeIRL	9iYoqT	908G2F	0650210	=ó@k;d/f.m
00.10	aUj71B	v42jbE	WLTZqP	908G2F	0710155	';,lk.;hj!

5.2   External Confidentiality Solution (ECS)

In the external environment, the plain part of the event log may cause data leakage. Therefore, the whole event log gets encrypted. Moreover, some additional attributes, which can lead to data leakage even in the encrypted form, are projected. In the following, our two-steps *ECS* is explained.

Encrypting the Plain Part. In this step, activities and resources are encrypted by a deterministic encryption method like AES. A deterministic encryption method must be used, because for discovering DFMs, differences should be preserved. Table 8 shows the result after encrypting activities and resources.

However, after encrypting, detecting "START" activities seem to be impossible, and without detecting them, finding traces becomes impossible. For identifying the "START" activities, we can go through the "Activity" ("Resource") and "Prev. Activity" ("Prev. Resource") columns, the activities (resources) which are appeared in the "Prev. Activity" ("Prev. Resource") column but not appeared in the "Activity" ("Resource") column are the "START" activities (resources).

Fortifying Encryption and/or Projecting Event Logs. As mentioned in Sect. 4, since resources are encrypted by a deterministic encryption method, and costs are encrypted by a homomorphic encryption method, which preserves differences, by comparison, one can find the minimum/maximum cost, which can be used as knowledge for extracting confidential or private information (e.g. name of resource). In order to decrease the effect of such analyses, fortifying encryption and/or projecting event logs could be done. Here, we project the costs which are indeed not necessary for the desired results.

6   Evaluation

We consider three evaluation criteria for the proposed approach, yet, at the same time, performance is also taken into account:

- *Ensuring Confidentiality:* As explained in Sect. 5, we can improve confidentiality by defining different environments and indicating a level of information which is accessible in each of these environments. In addition, using multiple encryption methods and our connector method for disassociating events from their cases provide high level of confidentiality with respect to the assumed background knowledge.
- *Reversibility:* When the keys and the value used for making times relative are given, both ICS and ECS are reversible, which means that transparency is addressed by the proposed approach.
- *Accuracy:* To show the accuracy of our approach, by a case study we illustrate that the results obtaining from the secure version of event logs are exactly the same as the results obtaining from the original event logs.

6.1   Correctness of the Approach

As can be seen in Fig. 3, from accuracy point of view, we need to show that the abstraction of the original event log is the same as the abstraction of the internal event log $(AEL = AEL')$ (rule (2)), and also the abstraction of the internal event log is the same as the abstraction of the external event log, which is encrypted $(AEL' = ECS^{-1}(AEL''))$ (rule (3)). To show that these relations are guaranteed to hold, we have implemented an interactive environment in Python and tested the approach on multiple event logs. In the following, we illustrate the results obtaining by applying the approach on "BPI challenge 2012".

$$AEL = AEL' \Rightarrow PM = PM' \wedge SN = SN' \tag{2}$$

$$AEL' = ECS^{-1}(AEL'') \Rightarrow PM' \approx ECS^{-1}(PM'') \wedge SN' \approx ECS^{-1}(SN'') \tag{3}$$

In the first step, EL', and EL'' were created. Then, to verify that AEL and AEL' are identical, we created a DFG from the original and the internal version of event log. Figure 6 shows the DFGs resulting from BPI challenge 2012 for the frequency threshold 2000. As one can see both DFGs are the same. Also, Fig. 7 shows the DFG resulting from BPI challenge 2012 for the same frequency threshold (2000) in the external environment. As can be seen, this DFG is also the same as the DFG from the EL and EL' (modulo renaming and layout differences), i.e., all the process discovery algorithms which are based on a DFG would lead to the same process models in the different environments.

In order to demonstrate that the causality based social networks in the secure environments are the same as the actual social networks from the original event log, we have made the real-handover from the original and internal version of event log for BPI challenge 2012. Figure 8 shows the networks for the real causal

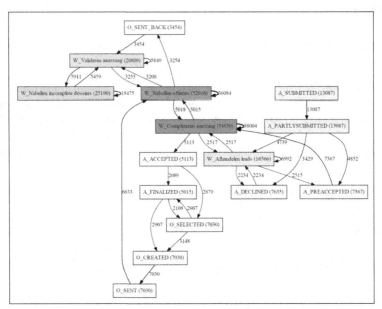

(a) The DFG from the original event log for the frequency threshold 2000.

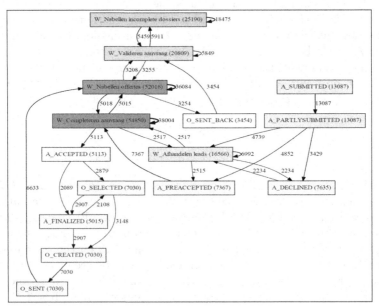

(b) The DFG from the internal event log for the frequency threshold 2000.

Fig. 6. Comparing the *DFG* from the *EL* with the *DFG* from the *EL'* for BPI challenge 2012: both graphs are identical, only layouts are different.

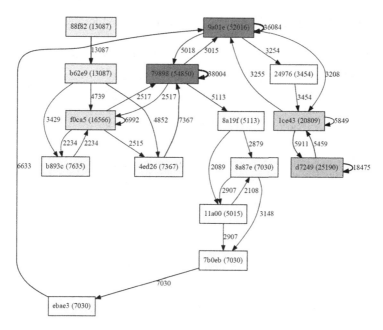

Fig. 7. The DFG from the external event log (BPI challenge 2012) for the frequency threshold 2000.

dependency threshold 0.5 and the frequency threshold 50. The networks are exactly the same. It is obvious that the network from the external version of event log must be the same (while resources are encrypted). Nevertheless, in Fig. 9, we have zoomed in the highlighted parts of Fig. 8 for the networks resulting from the internal and external environment (the same thresholds were applied), and relations are the same except the fact that resources in the external environment are encrypted. As can be seen in Fig. 9 all the relations of the resource "11201" are the same[1].

6.2 Performance

To demonstrate performance of the approach, we apply it on several benchmarking [4] and real-life event logs. Table 9 shows specifications of the used event logs. "BPI Challenge 2012" and "BPI Challenge 2017" are used to evaluate the performance when social networks are discovered, and the benchmarking event logs are used to evaluate the performance of the control-flow discovery.

Figure 10 shows how the control-flow discovery scales when using the benchmarking event logs and increasing the number of events exponentially, and Fig. 11 shows the performance of social network discovery when the approach is applied on the two real-life event logs with different scales. All runtimes are in

---

[1] It has 11 relations with the resources "112", "11000", "11189", "10913", "10861", "10909", "11181", "11180", "11119", "11203", and "11201".

Table 9. The specifications of the event logs used for evaluation

Event Log	Cases	Events	Variants	Activities	Resources
Choice Loop 1000	1000	7178	436	81	-
Choice Loop 10000	10000	70659	3202	81	-
Choice Loop 100000	100000	706598	21643	81	-
Sequence Loop 1000	1000	40783	1000	80	-
Sequence Loop 10000	10000	407791	9985	80	-
Sequence Loop 100000	100000	4078819	98821	80	-
BPI Challenge 2012	13087	262200	4366	24	69
BPI Challenge 2017	31509	561671	4047	26	145

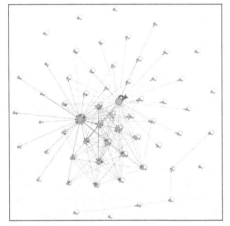 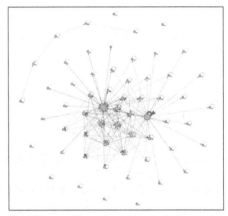

(a) The real-handover network from the original event log for the real causal dependency threshold 0.5 and the frequency threshold 50.

(b) The real-handover network from the internal event log for the real causal dependency threshold 0.5 and the frequency threshold 50.

Fig. 8. Comparing the real-handover networks resulting from BPI challenge 2012: both networks are identical.

milliseconds and have been tested using an Intel i7 Processor with 1.8 GHz and 16 GB RAM.

In Fig. 10, the darker bars show the execution time for discovering the DFG from the original event logs, and the lighter bars show the execution time for discovering the DFG from the secure event logs. One can see a linear increase of the runtime in milliseconds when adding choices or loops. In addition, as can be seen in Fig. 11, when the metric is real-handover, the execution time for discovering social networks is higher, since the real causal dependencies between subsequent resources need to be verified.

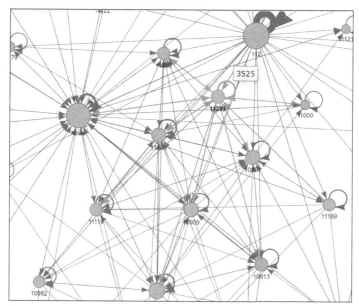

(a) The relations in the real-handover network from the internal event log for the real causal dependency threshold 0.5 and the frequency threshold 50.

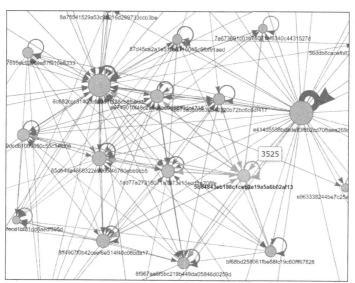

(b) The relations in the real-handover network from the external event log for the real causal dependency threshold 0.5 and the frequency threshold 50.

Fig. 9. Comparing the relations of resource "11201" in the real-handover networks resulting from EL' and EL'' for BPI challenge 2012

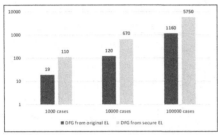
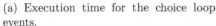
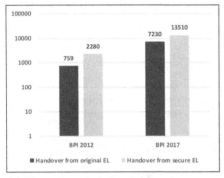

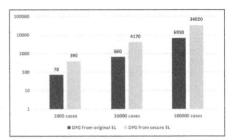
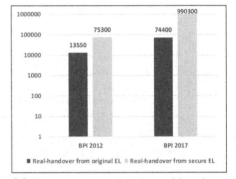

(a) Execution time for the choice loop events.

(b) Execution time for the sequence loop events.

Fig. 10. The execution time of the control-flow discovery when using the benchmaking event logs.

(a) Execution time for the handover networks.

(b) Execution time for the real-handover networks.

Fig. 11. The execution time of the social network discovery when using the real-life event logs.

7   Conclusions

This paper presented a novel approach to ensure confidentiality in process mining when the desired results are models. We demonstrated that confidentiality in process mining cannot be achieved by only encrypting an event log. We outlined the little related work, most of which use just encryption, and explained the weaknesses of following this approach. The new approach is introduced since there always exists a trade-off between confidentiality and data utility. Therefore, we reasoned backwards from the desired results and how they can be obtained with as little data as possible.

Here, process models and social networks were considered as the desired results, and the confidentiality solutions presented in the context of a framework that can be extended to include other forms of process mining, i.e., different ICS and ECS could be explored for different process mining activities. Moreover, the proposed framework could be utilized in cross-organizational context such that each environment could cover specific constraints and authorizations of a party. In this paper, we focused on causality based social networks, and in the future other metrics could be explored. Moreover, in the future, a measure for confidentiality could be defined so that the effectiveness of different solutions in this research area could be quantified and compared.

We have utilized a new method named "connector", which can be employed in any situation where we need to store associations securely. For evaluating the proposed approach, we have implemented an interactive environment in Python, and a real-life log was used as the case study.

Acknowledgment. We thank the Alexander von Humboldt (AvH) Stiftung for supporting our research interactions.

References

1. van der Aalst, W.M.P.: Business process management: a comprehensive survey. ISRN Softw. Eng. **2013**, 1–37 (2013)
2. van der Aalst, W.M.P.: Process Mining - Data Science in Action, Second edn. Springer, Heidelberg (2016). https://doi.org/10.1007/978-3-662-49851-4
3. van der Aalst, W.M.P.: Responsible data science: using event data in a "people friendly" manner. In: Hammoudi, S., Maciaszek, L.A., Missikoff, M.M., Camp, O., Cordeiro, J. (eds.) ICEIS 2016. LNBIP, vol. 291, pp. 3–28. Springer, Cham (2017). https://doi.org/10.1007/978-3-319-62386-3_1
4. van der Aalst, W.M.P.: Benchmarking logs to test scalability of process discovery algorithms. Eindhoven University of Technology (2017). https://data.4tu.nl/repository/uuid:1cc41f8a-3557-499a-8b34-880c1251bd6e. Accessed 01 Apr 2018
5. van der Aalst, W.M.P.: Process discovery from event data: relating models and logs through abstractions. Wiley Interdiscip. Rev.: Data Mining Knowl. Discov. **8**(3), e1244 (2018)
6. van der Aalst, W., et al.: Process mining manifesto. In: Daniel, F., Barkaoui, K., Dustdar, S. (eds.) BPM 2011. LNBIP, vol. 99, pp. 169–194. Springer, Heidelberg (2012). https://doi.org/10.1007/978-3-642-28108-2_19
7. van der Aalst, W.M.P., Adriansyah, A., van Dongen, B.: Replaying history on process models for conformance checking and performance analysis. Wiley Interdiscip. Rev.: Data Mining Knowl. Discov. **2**(2), 182–192 (2012)
8. van der Aalst, W.M.P., Bichler, M., Heinzl, A.: Responsible data science. Bus. Inf. Syst. Eng. **59**(5), 311–313 (2017)
9. van der Aalst, W.M.P., Reijers, H.A., Song, M.: Discovering social networks from event logs. Comput. Support. Coop. Work (CSCW) **14**(6), 549–593 (2005)
10. Accorsi, R., Stocker, T., Müller, G.: On the exploitation of process mining for security audits: the process discovery case. In: Proceedings of the 28th Annual ACM Symposium on Applied Computing, pp. 1462–1468. ACM (2013)
11. Bellare, M., Rogaway, P.: Introduction to modern cryptography. UCSD CSE **207**, 207 (2005)
12. Burattin, A., Conti, M., Turato, D.: Toward an anonymous process mining. In: 2015 3rd International Conference on Future Internet of Things and Cloud (FiCloud), pp. 58–63. IEEE (2015)
13. Daemen, J., Rijmen, V.: The design of Rijndael: AES-the advanced encryption standard. Springer, Heidelberg (2013)

14. Fahrenkrog-Petersen, S.A., van der Aa, H., Weidlich, M.: PRETSA: event log sanitization for privacy-aware process discovery. In: International Conference on Process Mining, ICPM 2019, Aachen, Germany, 24–26 June 2019, pp. 1–8 (2019)

15. Kapoor, V., Poncelet, P., Trousset, F., Teisseire, M.: Privacy preserving sequential pattern mining in distributed databases. In: Proceedings of the 15th ACM International Conference on Information and Knowledge Management, pp. 758–767. ACM (2006)

16. Katz, J., Menezes, A.J., Van Oorschot, P.C., Vanstone, S.A.: Handbook of Applied Cryptography. CRC Press, Boca Raton (1996)

17. Kleinberg, J.M.: Challenges in mining social network data: processes, privacy, and paradoxes. In: Proceedings of the 13th ACM SIGKDD International Conference on Knowledge Discovery and Data Mining, pp. 4–5. ACM (2007)

18. Leemans, M., van der Aalst, W.M.P., van den Brand, M.G.: Hierarchical performance analysis for process mining. In: Proceedings of the 2018 International Conference on Software and System Process, pp. 96–105. ACM (2018)

19. Leemans, S.J.J., Fahland, D., van der Aalst, W.M.P.: Scalable process discovery and conformance checking. Softw. Syst. Model. 17(2), 599–631 (2016). https://doi.org/10.1007/s10270-016-0545-x

20. Liu, C., Duan, H., Qingtian, Z., Zhou, M., Lu, F., Cheng, J.: Towards comprehensive support for privacy preservation cross-organization business process mining. IEEE Trans. Serv. Comput. 1, 1–1 (2016)

21. Ma, C.Y., Yau, D.K., Yip, N.K., Rao, N.S.: Privacy vulnerability of published anonymous mobility traces. IEEE/ACM Trans. Netw. (TON) 21(3), 720–733 (2013)

22. Mannhardt, F., de Leoni, M., Reijers, H.A., van der Aalst, W.M.P., Toussaint, P.J.: Guided process discovery-a pattern-based approach. Inf. Syst. 76, 1–18 (2018)

23. Mannhardt, F., Petersen, S.A., Oliveira, M.F.: Privacy challenges for process mining in human-centered industrial environments. In: 2018 14th International Conference on Intelligent Environments (IE), pp. 64–71. IEEE (2018)

24. Paillier, P.: Public-key cryptosystems based on composite degree residuosity classes. In: Stern, J. (ed.) EUROCRYPT 1999. LNCS, vol. 1592, pp. 223–238. Springer, Heidelberg (1999). https://doi.org/10.1007/3-540-48910-X_16

25. Pourbafrani, M., van Zelst, S.J., van der Aalst, W.M.P.: Scenario-based prediction of business processes using system dynamics. In: Panetto, H., Debruyne, C., Hepp, M., Lewis, D., Ardagna, C.A., Meersman, R. (eds.) OTM 2019. LNCS, vol. 11877, pp. 422–439. Springer, Cham (2019). https://doi.org/10.1007/978-3-030-33246-4_27

26. Rafiei, M., van der Aalst, W.M.P.: Mining roles from event logs while preserving privacy. In: Di Francescomarino, C., Dijkman, R., Zdun, U. (eds.) BPM 2019. LNBIP, vol. 362, pp. 676–689. Springer, Cham (2019). https://doi.org/10.1007/978-3-030-37453-2_54

27. Rafiei, M., von Waldthausen, L., van der Aalst, W.M.P.: Ensuring confidentiality in process mining. In: Proceedings of the 8th International Symposium on Data-driven Process Discovery and Analysis (SIMPDA 2018), Seville, Spain, 13–14 December 2018, pp. 3–17 (2018). http://ceur-ws.org/Vol-2270/paper1.pdf

28. Fani Sani, M., van Zelst, S.J., van der Aalst, W.M.P.: Repairing outlier behaviour in event logs. In: Abramowicz, W., Paschke, A. (eds.) BIS 2018. LNBIP, vol. 320, pp. 115–131. Springer, Cham (2018). https://doi.org/10.1007/978-3-319-93931-5_9

29. Tillem, G., Erkin, Z., Lagendijk, R.L.: Privacy-preserving alpha algorithm for software analysis. In: 37th WIC Symposium on Information Theory in the Benelux/6th WIC/IEEE SP Symposium on Information Theory and Signal Processing in the Benelux (2016)
30. Wasserman, S., Faust, K.: Social Network Analysis: Methods and Applications, vol. 8. Cambridge University Press, Cambridge (1994)
31. Zhan, J.Z., Chang, L., Matwin, S.: Privacy-preserving collaborative sequential pattern mining. Technical report, Ottawa Univ (Ontario) School of Information Technology (2004)

Author Index

Printed in the United States
By Bookmasters